Regulatory and Pharmacological Basis of Ayurvedic Formulations

Regulatory and Pharmacological Basis of Ayurvedic Formulations

Amritpal Singh Saroya
Herbal Consultant
Punjab
India

CRC Press
Taylor & Francis Group
Boca Raton London New York

CRC Press is an imprint of the
Taylor & Francis Group, an **informa** business

A SCIENCE PUBLISHERS BOOK

CRC Press
Taylor & Francis Group
6000 Broken Sound Parkway NW, Suite 300
Boca Raton, FL 33487-2742

First issued in paperback 2020

© 2016 by Taylor & Francis Group, LLC
CRC Press is an imprint of Taylor & Francis Group, an Informa business

No claim to original U.S. Government works

ISBN-13: 978-1-4987-5094-3 (hbk)
ISBN-13: 978-0-367-78292-4 (pbk)

This book contains information obtained from authentic and highly regarded sources. Reasonable efforts have been made to publish reliable data and information, but the author and publisher cannot assume responsibility for the validity of all materials or the consequences of their use. The authors and publishers have attempted to trace the copyright holders of all material reproduced in this publication and apologize to copyright holders if permission to publish in this form has not been obtained. If any copyright material has not been acknowledged please write and let us know so we may rectify in any future reprint.

Except as permitted under U.S. Copyright Law, no part of this book may be reprinted, reproduced, transmitted, or utilized in any form by any electronic, mechanical, or other means, now known or hereafter invented, including photocopying, microfilming, and recording, or in any information storage or retrieval system, without written permission from the publishers.

For permission to photocopy or use material electronically from this work, please access www.copyright.com (http://www.copyright.com/) or contact the Copyright Clearance Center, Inc. (CCC), 222 Rosewood Drive, Danvers, MA 01923, 978-750-8400. CCC is a not-for-profit organization that provides licenses and registration for a variety of users. For organizations that have been granted a photocopy license by the CCC, a separate system of payment has been arranged.

Trademark Notice: Product or corporate names may be trademarks or registered trademarks, and are used only for identification and explanation without intent to infringe.

Library of Congress Cataloging-in-Publication Data

Names: Amritpal Singh, 1971- , author.
Title: Regulatory and pharmacological basis of ayurvedic formulations / Amritpal Singh Saroya.
Description: Boca Raton : Taylor & Francis, 2017. | Includes bibliographical references and index.
Identifiers: LCCN 2016010614 | ISBN 9781498750943 (hardcover : alk. paper)
Subjects: | MESH: Medicine, Ayurvedic | Materia Medica--pharmacology | Drug and Narcotic Control | Drug Compounding | Chemistry, Pharmaceutical
Classification: LCC R733 | NLM WB 55.A9 | DDC 615.5/38--dc23
LC record available at http://lccn.loc.gov/2016010614

Visit the Taylor & Francis Web site at
http://www.taylorandfrancis.com

and the CRC Press Web site at
http://www.crcpress.com

PREFACE

We feel happy and satisfied in presenting this book titled *Regulatory and Pharmacological Issues of Ayurvedic Formulations* before the pharmacy fraternity. Research in Ayurvedic medicine is a buzzword and several national and international companies of repute are investing funds for novel and cost-effective drug-discovery from natural sources. Ayurvedic Pharmaceutical Sciences is an upcoming discipline and is expected to play a pivotal role in future healthcare industry.

The idea of writing a book on such a vast topic was an on-going and vigorous effort. Standard and practical books on the pharmaceutical value of Ayurvedic Medicine have been lacking. Standard books incorporating information on standardisation, pre-clinical (animal) and clinical studies and toxicity data of widely used Ayurvedic formulations have been a missing link.

The book has two distinct sections (evident from the title)

Regulatory: This section deals with regulatory affairs related to Ayurvedic formulations. The introductory chapter throws light on regulatory affairs. Succeeding chapters are correlated to Ayurvedic pharmacopoeia, Drug and Cosmetic Act, Licensing and GMP issues related to the Ayurvedic drug industry, pharmacovigilance, genotoxic potential of Ayurvedic formulations and so on. This section shall serve as benchmark for the Ayurvedic drug industry.

Pharmacological: Section B is dedicated to pharmacological investigations done on selected Ayurvedic formulations. This section highlights pharmaceuticals and standardisation issues of Ayurvedic formulations. Animal, toxicological and clinical studies have been arranged under one roof thereby making the section multi-dimensional. This section shall fulfil requirement of students, researchers and academicians.

I sincerely hope that this book will be accepted by the Ayurvedic and pharmacy fraternity.

Dr. Amritpal Singh Saroya
MD (Ayurveda); MSc (Medicinal Plants)

CONTENTS

Preface   v

PART A: AYURVEDA AND REGULATORY AFFAIRS

1. Introduction to Regulatory Affairs   3
2. Introduction to Ayurvedic Pharmacopoeia of India   8
3. Drugs and Cosmetic Act and Ayurvedic Drugs   14
4. Ayurvedic Drug Manufacturing License   22
5. Good Manufacturing Practices for ASU Medicines (Schedule-T)   26
6. Ayurvedic Drug Industry   37
7. Pharmacovigilance of Ayurvedic Drugs   41
8. Heavy Metal Content of Ayurvedic Formulations   46
9. Genotoxic Potential of Ayurvedic Formulations   54
10. Aristolochic Acid Distribution in Ayurvedic Formulations   55
11. Clinical Trials in Ayurveda   58
12. National Ayush Mission   64
13. Ayurvedic Pharmacy Education   67
14. Patent and IPR Issues of Ayurvedic Formulations   70
15. Ayurvedic Pharmacoepidemiology   75
16. Voluntary Certification Scheme for Ayush Products   77
17. Ayurvedic Databases   88
18. Medicinal Plants Division of Indian Council for Medical Research   90
Bibliography (Part A)   93

PART B: PHARMACOLOGICAL INVESTIGATIONS ON AYURVEDIC FORMULATIONS

A-Z of Standardisation, Pre-Clinical, Clinical and Toxicological Data	99
Bibliography (Part B)	192
Notifications	213
Forms	224
Tables	227
Lists	240
Annexures	251
Index	257

PART A
AYURVEDA AND REGULATORY AFFAIRS

CHAPTER 1
INTRODUCTION TO REGULATORY AFFAIRS

1.1. DEFINITION

Regulatory affairs (RA) is also known as government affairs. A regulatory affair is a profession within regulated industries, such as pharmaceuticals, medical devices, energy, banking, telecom, etc. Regulatory affairs have a very specific meaning within the healthcare industries including pharmaceuticals, medical devices, biologic and functional foods.

1.2. IMPORTANCE OF REGULATORY AFFAIRS

Regulatory affairs are a crucial function in the pharmaceutical industry. Industries like pharmaceuticals, biologics, food and medical equipment can be uncertain if the new products and methods are not tested and checked for functionality with great vigilance before being publicised.

The first and foremost factor for the pharmaceutical sector has been always in the efficacy and cost-effectiveness of the drug contributing to the success in the market. Therefore, effective administration and superintendence of regulatory affairs actions plays a crucial role in the economy of the corporation.

The regulatory affairs professional is the only person who is completely responsible for such activities developing distinct rejoinders to regulatory authorities who want an organization to hold products in compliance and retain satisfactory data in support of applications that have been made for the registration of the products.

1.3. TRADITIONAL HERBAL MEDICINES AND HUMAN HEALTH

Herbal medicines which have formed the basis of health care throughout the world since the earliest days of mankind are still widely used, and have considerable importance in international trade. Recognition of their clinical, pharmaceutical and economic value is still growing, although this varies widely between countries.

Medicinal plants are important for pharmacological research and drug development, not only when plant constituents are used directly as therapeutic agents, but also as starting materials for the synthesis of drugs or as models of pharmacologically active compounds. Regulation of exploitation and exportation is therefore essential, together with international cooperation and coordination for the conservation of these plants so as to ensure their availability in the future.

The United Nations Convention on Biological Diversity states that the conservation and sustainable use of biological diversity is of critical importance for meeting the food, health and other needs of the growing world population. For this purpose, access to and sharing of both genetic resources and technologies is essential.

Legislative controls in respect to medicinal plants have not evolved around a structured control model. There are different ways in which countries define medicinal plants or herbs or products derived from them. Countries have adopted various approaches to licensing, dispensing, manufacturing and trading to ensure the safety, quality and efficacy of the medicinal plants.

Despite the use of herbal medicines over many centuries, only a relatively small number of plant species have been studied for possible medical applications. However, safety and efficacy data are available for an even smaller number of plants, their extracts and active ingredients and preparations containing them.

1.4. REGULATION AND REGISTRATION OF HERBAL MEDICINES

The legal situation regarding herbal preparations varies from country to country. In sum, phytomedicines are well-established, whereas in others they are regarded as food and therapeutic claims are not allowed. Developing countries, however, often have a great number of traditionally used herbal medicines and much folk-knowledge about them, but have hardly any legislative criteria to establish these traditionally used herbal medicines as part of the drug legislation.

For the classification of herbal or traditional medicinal products, factors applied in regulatory systems include: description in a pharmacopoeia monograph, prescription status, the claim of a therapeutic effect, scheduled or regulated ingredients or substances, or periods of use. Some countries draw a distinction between "officially approved" products and "officially recognized" products, by which the latter products can be marketed without scientific assessment by the authority.

The various legislative approaches for herbal medicines fall into one or other of the following categories:

1. Same regulatory requirements for all products;
2. Same regulatory requirements for all products, with certain types of evidence, not required for herbal/traditional medicines;
3. Exemption from all regulatory requirements for herbal/traditional medicines;
4. Exemption from all regulatory requirements for herbal/traditional medicines concerning registration or marketing authorization;
5. Herbal/traditional medicines subject to all regulatory requirements; and
6. Herbal/traditional medicines subject to regulatory requirements concerning registration or marketing authorization.

Where herbal medicines and related products are neither registered nor controlled by regulatory bodies, a special licensing system is needed which would enable health authorities to screen the constituents, demand proof of quality before marketing, ensure correct and safe use, and also to oblige license holders to report suspected adverse reactions within a post-marketing surveillance system.

1.5. THE WHO TRADITIONAL MEDICINE PROGRAMME

The World Health Assembly (WHA) has adopted a number of resolutions drawing attention to the fact that a large section of the population in many developing countries still relies on traditional medicine, and that the workforce represented by traditional practitioners is a potentially important resource for primary health care. In 1978, the Declaration of Alma-Ata recommended, inter alia, the inclusion of proven traditional remedies into national drug policies and regulatory measures.

The policy of the World Health Organization regarding traditional medicine was presented in the Director General's report on Traditional Medicine and Modern Health Care to the Forty-fourth World Health Assembly 1991, which stated that "WHO collaborated with its Member States in the review of national policies, legislation and decisions on the nature and extent of the use of traditional medicine in their health systems."

Herbal medicines have been included in the International Conference on Drug Regulatory Authorities (ICDRA) since the Fourth Conference in 1986. A WHO consultation in Munich, Germany, June 1991, drafted Guidelines for the Assessment of Herbal Medicines which was adopted for general use by the Sixth ICDRA in Ottawa, October 1991. These guidelines (WHO/TRM/91.4) define basic criteria for the evaluation of quality, safety and efficacy of herbal medicines to assist national regulatory authorities, scientific organizations, and manufacturers to undertake an assessment of the, of submissions and/or the dossiers in respect to such products.

In 1994, the WHO Regional Office for the Eastern Mediterranean published Guidelines for Formulation of National Policy on Herbal Medicines. A WHO consultation on "WHO Monographs on Selected Medicinal Plants" took place in Munich, Germany 1996. After discussion and review, 28 monographs were adopted. The purpose of this document is: to provide scientific information on the safety, efficacy and quality control of widely used medicinal plants; to facilitate the proper use of herbal medicines; to provide models for Member States to develop their own monographs on these and additional herbal medicines; and to facilitate information exchange. The 28 monographs were presented at the Eighth ICDRA meeting in Bahrain, November 1996. Another 32 monographs are being prepared.

1.6. REGULATORY ASPECTS OF AYURVEDA

Globally, there have been concerted efforts to monitor quality and regulate the growing business of herbal drugs and traditional medicine. Health authorities and governments of various nations have taken an active interest in providing standardized botanical medications. Government of India has also plunged into this opportunity and initiated some regulations in this sector.

To ensure and enhance the quality of ASU medicines, the Government of India has notified Good Manufacturing Practices (GMP) under Schedule 'T' of the Drugs and Cosmetics Act, 1940 which also ensures raw materials used in the manufacture of drugs are authentic, of prescribed quality and are free from contamination.

The guidelines for Good Agricultural Practices (GAP) seek to lay down a cultivation program designed to ensure optimal yield in terms of both quality and quantity of any crop intended for health purposes. It puts forth a standard for production of raw material that goes into the making of the ASU medicines and standardizes the production processes from farm to factory.

As a matter of fact, it can be seen that there is a direct correlation between efficacy of a herbal drug with the quality of raw material used and the process of manufacturing. It is of paramount importance that no compromise is made on quality. To ensure that the quality of raw material is being watched by following GAP and GACP (for manufacturing and marketing the prepared drugs), the government has formulated the Drugs and Cosmetics Act, 1940. It is an act to regulate the import, manufacture, distribution and sale of drugs and cosmetics.

This act was basically initiated for chemical drugs, but later in the year 1969, a separate chapter relating to Ayurveda, Siddha and Unani drugs was inserted by act 13 of 1964. The laws are partly same as those for conventional pharmaceuticals. Later on, this act was again modified with some substitutions in the year 1983, 1987, 1994 and 2002. The schedules and rules pertaining to Ayurveda, Siddha and Unani systems in the act are:

Schedules

- First Schedule substituted by act 13 of 1964 came into force w.e.f. 1.2.1969. The schedule lists the standard Indian pharmacopoeias to be followed for manufacturing Ayurveda, Siddha and Unani drugs. About 57 books of Ayurveda (with insertions in 1987, 1994, 2002), 29 of Siddha (1987), 13 of Unani Tibb systems are listed.
- Second Schedule came into force w.e.f. 15.09.64. It states the standard to be complied with for manufacturing drugs. (Subs. by Notification No. G.S.R. 885, dated the 4th August, 1973, Gazette of India, Pt. II, s. 3(i), p. 1643.)
- SCHEDULE-E (1): List of poisonous substances under the Ayurvedic (including Siddha) and Unani Systems of Medicine (Added by Notification No. 1-23/67-D dt. 2.2.1970) differentiated into vegetable, animal and mineral origin.
- SCHEDULE T: Good Manufacturing Practices (GMP) for Ayurvedic, Siddha and Unani medicines. (Ins. by G.S.R. 561 (E) dt. 23-06-2000 and subs. by G.S.R. 198(E), dt. 7.3.2003.) Under Schedule "T" of the Drugs and Cosmetics act 1940, the government has made it mandatory for all manufacturing units to adhere to GMP.

Rules

- Rules: Part XVI (Parts XVI, XVII and XVII added by S.O. 642, dt. the 2.2.1970 (w.e.f. 21.2.1970)). Manufacture for sale of Ayurvedic (including Siddha) or Unani drugs. It concerns how to acquire licenses, loans for establishing a unit and also the identification of raw materials and their purity.
- Part XVIA: Approval of institutions for carrying out tests on Ayurvedic, Siddha and Unani drugs and raw materials used in their manufacture on behalf of licenses to manufacture for sale of Ayurvedic, Siddha and Unani drugs. (Ins. by G.S.R. 701(E), dt. 27.7.2001 and subs. by G.S.R. 73(E), dt. 31.01.2003.)
- Part XVII: Labelling, packing and limit of alcohol in Ayurvedic (including Siddha) or Unani drugs. (Subs. by G.S.R. 904(E), dt. 2.11.1992.)
- PART XVIII: Government analysts and inspectors for Ayurvedic (including Siddha) or Unani drugs.
- PART XIX: Standards of Ayurvedic, Siddha or Unani. (Ins. by G.S.R. 519(E), dt. 26.6.1995.)

CHAPTER 2
INTRODUCTION TO AYURVEDIC PHARMACOPOEIA OF INDIA

2.1. DEFINITIONS

Pharmacopoeia: Pharmacopoeia literally means "drug-making". Pharmacopoeia is a book describing drugs, chemicals and medicinal preparations; this pertains especially to one issued by an officially recognized authority and serving as a standard.

Ayurvedic Pharmacopoeia of India (API): The Ayurvedic Pharmacopoeia of India is a legal document of standards for the quality of Ayurvedic drugs and substances included therein (under the Drugs and Cosmetic Act, 1940). The API is published by the Department of Ayurveda, Yoga & Naturopathy, Unani, Siddha and Homoeopathy (AYUSH), Ministry of Health and Family Welfare, Government of India.

Ayurvedic Formulary of India (AFI): The scattered information on various formulations in classical Ayurvedic books has been compiled in such a way to make it suitable to develop pharmacopoeial standards and also to meet the requirements of the Drugs and Cosmetics Act.

Pharmacopoeial Laboratory for Indian Medicine (PLIM): Pharmacopoeial Laboratory for Indian Medicine is a subordinate office of the Ministry of Health & Family Welfare (Deptt. of AYUSH), Govt. of India. PLIM is a Standards Setting cum Drugs Testing Laboratory at National Level for Indian Medicines which include drugs of Ayurveda, Unani and Siddha systems.

Essential Drug List (EDL): To overcome the problem lack of availability of Ayush medicines in the public health system and facilitate the state and central authorities for smooth procurement of medicines, the department of Ayush has issued the essential drugs list (EDL) of Ayurveda, Siddha, Unani and Homoeopathy medicines. There are 277 essential medicines in the EDL of Ayurveda, 257 medicines in Homoeopathy, 302 medicines in Siddha and 288 essential medicines in the Unani system of medicine.

ASUTAB: Ayurveda, Siddha and Unani Drugs Technical Advisory Board. Department of Ayush has constituted five sub-committees under ASUDTAB.

- The first Sub Committee will examine Schedule 'Z' and other relevant notifications.
- The second Sub Committee is formed to evaluate the proposed Retail Sale License for ASU Drugs.
- The third Sub Committee reviews the Model Laboratory Practices for testing of ASU drugs.
- The fourth Sub Committee is formed to amend the First Schedule of Drugs & Cosmetics Act, 1940 for the list of Authoritative ASU Books.
- The fifth Sub Committee examines the shelf life of ASU Medicines.

ADUDCC: ASU Drug Consultative Committee.
ADMA: Ayurveda Drug Manufacturers Association.
AMMOI: Ayurvedic Medicine Manufacturers Organization of India.

2.2. PHARMACOPOEIA COMMISSION FOR INDIAN SYSTEM OF MEDICINE

To lay down Pharmacopoeia Standards, Govt. of India has constituted three Pharmacopoeia Committees viz;

1. Ayurvedic Pharmacopoeia Committee,
2. Siddha Pharmacopoeia Committee &
3. Unani Pharmacopoeia Committee.

The purpose of the above three is to lay down Pharmacopoeial Standards for Ayurvedic, Siddha & Unani drugs. Pharmacopoeial Laboratory for Indian Medicines, Ghaziabad is the only laboratory to develop pharmacopoeial standards at present. Govt. of India has published standards on 540 single drugs and 101 compound formulations. Still, there are about 2000 single drugs and about an equal number of compounded formulations where standards are yet to be developed.

Quality standards of Ayurveda, Siddha and Unani medicines and their periodic up-to-date are a top priority of the Department to cater the needs of the consumers. The popularity and demand for ASU medicines are increasing rapidly. Therefore, it is necessary to upgrade pharmacopoeia committees to a Pharmacopoeia Commission of Indian Medicine and PLIM, Ghaziabad will be the infrastructure support of the Commission. The traditional medicines and natural/herbal products have a global demand. Therefore, the proposed Commission will work on the lines of other pharmacopoeia commissions of the world like US Pharmacopoeia Commission, EU Pharmacopoeia Commission and British Pharmacopoeia Commission.

The standards require to be upgraded regularly because the safety requirements tend to get more and more stringent. The Pharmacopoeia is necessarily a dynamic concern and updating of standards is an ongoing exercise and it cannot be done by anyone other than an independent body, which alone can ensure promptness and in-depth studies – essential features of this exercise.

The emphasis these days has to be more on limiting undesirable impurities in bulk drugs and formulations than merely ensuring minimum standards of purity. There is fast progress in the evolution of more sophisticated methods of analysis. These rapid developments should be reflected in successive editions of pharmacopoeia which must perforce be brought out more frequently than is done at present. A well organized independent body may be able to assure this, and furthermore, the standards embodied in monographs can be fully unbiased.

The Pharmacopoeia Commission for Indian Medicine will cater to the need of ASU Pharmacopoeia Committees and PLIM. This will be a fully autonomous organization registered under the Society Act. The organizational set up of the Commission has been proposed to be a pure scientific body having full autonomy to recruit and co-ordinate scientists, utilize funds more efficiently and also generate resources to meet the targets of the commission.

For allopathic medicines, Indian Pharmacopoeia Commission was set up in 2006. Learning the lessons from other international pharmacopoeia commissions as well as an Indian Pharmacopoeia Commission, the proposed Pharmacopoeia Commission for Indian Medicine has taken the best from all other commissions. The objects of the commission are as follows.

2.2.1. Objectives

Pharmacopoeia Commission for Indian System of Medicine will work as a standardization and quality control laboratory at the national level and impart training on quality control and GMP to personnel from Manufacturer/Industries and Regulatory bodies.

1. Publication and revision of the Ayurvedic, Siddha and Unani Pharmacopoeia of India at suitable intervals and of such addenda or supplementary compendia during the intervening periods as may be deemed necessary; releasing the publications for public use from a date when they are to become official.
2. Publication and revision of the Ayurvedic, Siddha and Unani Formulary of India at regular intervals with a view to make it a authentic source of information on the rational combination and use of medicines, including their method of preparation, therapeutic indications, adverse reactions, contraindications, drug-drug interactions and similar issues concerning Indian medicines for safe use in humans and animals and identification of ASU Formulation with a view to develop their quality standards and to ensure quality and safety of ASU medicines;
3. To nurture and promote awareness of quality in ASU drugs formulations and drug research on ASU products and publish regularly or at suitable intervals other related scientific information as authorised under the rules and procedures;
4. Exchange information and interact with expert committees of the World Health Organization and other international bodies with a view to harmonise and develop the ASU pharmacopoeial standards so as to make them internationally acceptable;
5. Arranging studies either under its own auspices or through collaboration with other institutions to develop standards and quality specifications for identity, purity and strength of raw materials and compound formulations and to develop

SOPs for the process of manufacture included or to be included in the Ayurvedic, Siddha and Unani Pharmacopoeia/formulary and its addenda or supplementary compendia or other authorized publications;
6. Maintain a national repository of authentic reference raw materials used in the manufacture of Ayurveda, Siddha and Unani medicines for the purpose of reference and supply of reference standards to the stakeholders at a price;
7. Generate and maintain a repository of chemical reference marker compounds of the plants or other ingredients used in standardizing Ayurveda, Siddha and Unani medicines and supply them as reference standards to the stakeholders on price;
8. Furtherance of the provision of Chapter IVA of Drugs and Cosmetic Act & Rules there under related to ASU Drugs;
9. Acting as a coordinating center for analytical laboratories, industry and academia by encouraging the exchange of scientific and technical information and staff and by undertaking sponsored funded research as well as consultancy projects;
10. Organizing a national/international symposia, seminars, meetings and conferences in selected areas from time to time and to provide updated regular training to the regulatory authorities and stakeholders.

2.3. AYURVEDIC PHARMACOPOEIA OF INDIA (API)

2.3.1. The Members & Co-opted Members of Ayurvedic Pharmacopoeia Commission

A. Composition of Ayurvedic Pharmacopoeial Committee
B. Project Evaluation and Monitoring Committee
C. Screening Committee

2.3.2. Ayurvedic Pharmacopoeia Committee

The 1st APC was constituted in 1962 and has been working continuously since then. The term of the Committee remains for a period of 3 years from the date of its first meeting and the members hold office for that period.

- APC was functional at Dept. of AYUSH; however, it has been shifted to the Council in April 2006. Since then, Director General, CCRAS (erstwhile Director, CCRAS) is designated as Member Secretary.
- The existing Main Committee of APC (2009) with 23 (+13 co-opted) eminent members are chaired by Prof. S. S. Handa.
- Advisor (Ay.), Dept. of AYUSH, Ministry of Health & Family Welfare is the Vice Chairman.

For finalization of Pharmacopoeial work, following subcommittees are functioning under APC:

1. Formulary Subcommittee of APC (Ras Shastra/Bhaishjya kalpana-Ayurvedic Pharmacy),

2. Ayurveda Single Drugs Subcommittee (Single Drugs of Plants, Minerals, Metals & Animal origin),
3. Subcommittee of Pharmacogonosy,
4. Subcommittee of Photochemistry & Chemistry.

The functions of APC are as follows:

- To prepare Ayurvedic Pharmacopoeia of India (API) of single drugs (Part I) and compound formulations (Part II).
- To prescribe the working standards for raw materials as well as compound formulations including tests for identity, purity, strength and quality so as to ensure uniformity of the finished formulations.
- To develop and standardize method of preparations, dosage forms, toxicity profile, etc. of formulations.
- To identify methods and procedures for publication of the standards of all commonly used formulations of AFI in a phased manner.
- To provide all other information on Ayurvedic formulations regarding the distinguishing characteristics, methods of preparation, dosage, method of administration with various anupans or vehicles and their toxicity.
- To develop the quality standards, safety, efficacy profile of different parts of the plants as well as the inclusion of new plants as Ayurvedic drugs.
- Any other matter relating to the quality standards, shelf life, identification, new formulations, etc.
- To develop quality standards, safety, efficacy profile of Intermediates like extracts of plant drugs used in Ayurveda.

Publications

Ayurvedic Formulary of India

To bring uniformity among the manufacturers and to follow the same formula of ingredients in the same proportion, two parts of Ayurvedic Formulary of India have been published.

- In the Ayurvedic Formulary of India Part I and Part II there are 635 Formulations. Both parts are available in English and Hindi separately.
  - AFI Part I (1978) – 444 formulations
  - AFI Part II (2000) – 191 formulations
  - Part III with simultaneous descriptions in Hindi and English is in draft stage.

Ayurvedic Pharmacopoeia of India

A. Seven Volumes of Ayurvedic Pharmacopoeia Part, I have been published consisting of 540 monographs on single drugs.
B. Three Volumes of Ayurvedic Pharmacopoeia Part, II have been published consisting of 101 monographs on formulations.

2.4. AYUSH MARK

AYUSH MARK Quality Council of India has been engaged in the voluntary certification of quality of Ayurveda, Siddha Unani (ASU) products. Through this scheme drug manufacturers are awarded the quality seal to the products on the basis of third party evaluations of the quality, subject to fulfillment of the regulatory requirements. AYUSH Standard and AYUSH Premium Marks are awarded for products moving in the domestic and international market respectively. 146 ASU products are reported to have been awarded the AYUSH Premium Mark and 97 products the AYUSH Standard Mark.

CHAPTER 3
DRUGS AND COSMETIC ACT AND AYURVEDIC DRUGS

3.1. BASIC DEFINITIONS

A. **Drugs and Cosmetic Act,** 1940 is an act to regulate the import, manufacture, distribution and sale of drugs and cosmetics.
B. **"Ayurvedic, Siddha or Unani drugs"** includes all medicines intended for internal or external use for or in the diagnosis, treatment, mitigation or prevention of disease or disorder in human beings or animals, and manufactured exclusively in accordance with the formulate described in the authoritative books of Ayurvedic, Siddha and Unani Tibb system of medicine specified in the First Schedule.
C. **"Board"** means, in relation to the Ayurvedic, Siddha or Unani drugs, the Ayurvedic, Siddha and Unani Drugs Technical Advisory Board constituted under section 33C.
D. **"Cosmetic"** means any article intended to be rubbed, poured sprinkled or sprayed on, or introduced into, or otherwise applied to, the human body or any part thereof for cleansing, beautifying, promoting attractiveness, or altering the appearance, and includes any article intended for use as a component of cosmetic.
E. **"Government Analyst"** means, in relation to Ayurvedic, Siddha or Unani drugs, a Government Analyst appointed by the Central Government or a State Government under section 33F.
F. **"Inspector"** means, in relation to Ayurvedic, Siddha or Unani drugs, an Inspector appointed by the Central Government or a State Government under section 33G.
G. **"Patent or proprietary medicine"** means, in relation to Ayurvedic, Siddha or Unani Tibb systems of medicine all formulations containing only such ingredients mentioned in the formulate described in the authoritative books of Ayurveda, Siddha or Unani Tibb systems of medicine specified in the First Schedule, but does not include a medicine which is administered by parental route, and also a formulation included in the authoritative books as specified in clause (a). OK

Note: Section 5 (The Drugs Technical Advisory Board) and 7 (The Drugs Consultative Committee) does not to apply to Ayurvedic, Siddha or Unani drugs. – Nothing contained in sections 5 and 7 shall apply to Ayurvedic, Siddha or Unani drugs.

3.2. PROVISIONS OF DRUGS AND COSMETIC ACT, 1940 APPLICABLE TO ASU DRUGS

33C. Ayurvedic, Siddha and Unani Drugs Technical Advisory Board
33D. Ayurvedic, Siddha and Unani Drugs Consultative Committee
33E. Misbranded drugs
33EE. Adulterated drugs
33EEA. Spurious drugs
33EEB. Regulation of manufacture for sale of ASU drugs
33EEC. Prohibition of manufacture and sale of certain ASU drugs
33EED. Power of Central Government to prohibit manufacture, etc., of ASU drugs in public interest
33F. Government Analysts
33G. Inspectors
33H. Application of provisions of sections 22, 23, 24 and 25
33-I. Penalty for manufacture, sale, etc., of ASU drugs in contravention of this Chapter
33J. Penalty for subsequent offenses
33K. Confiscation
33L. Application of provisions to Government departments
33M. Cognizance of offenses
33N. Power of Central Government to make rules
33O. Power to amend First Schedule

3.3. DETAILS OF PROVISIONS OF DRUGS AND COSMETIC ACT, 1940 APPLICABLE TO ASU DRUGS (CHAPTER IVA)

33C.12 [Ayurvedic, Siddha and Unani Drugs Technical Advisory Board].—
1) The Central Government shall, by notification in the Official Gazette and with effect from such date as may be specified therein, constitute a Board (to be called the 12 [Ayurvedic, Siddha and Unani Drugs Technical Advisory Board]) to advise the Central Government and the State Governments on technical matters arising out of this Chapter and to carry out the other functions assigned to it by this Chapter.

2) The Board shall consist of the following members, namely:—
   i) The Director General of Health Services, ex officio;
   ii) The Drugs Controller, India, ex officio;
   iii) The principal officer dealing with Indian systems of medicine in the Ministry of Health, ex officio;
   iv) The Director of the Central Drugs Laboratory, Calcutta, ex officio;
   v) One person holding the appointment of Government Analyst under section 33F, to be nominated by the Central Government;
   vi) One Pharmacognocist to be nominated by the Central Government;

vii) One Phyto-chemist to be nominated by the Central Government;
viii) Four persons to be nominated by the Central Government, two from amongst the members of the Ayurvedic Pharmacopoeia Committee, one from amongst the members of the Unani Pharmacopoeia Committee and one from amongst the members of the Siddha Pharmacopoeia Committee;
ix) One teacher in Dravyaguna and Bhaishajya Kalpana, to be nominated by the Central Government;
x) one teacher in ILM-UL-ADVIA and TAKLIS-WA-DAWA-SAZI, to be nominated by the Central Government;
xi) One teacher in Gunapadam, to be nominated by the Central Government;
xii) Three persons, one each to represent the Ayurvedic, Siddha and Unani drug industry, to be nominated by the Central Government;
xiii) Three persons, one each from among the practitioners of Ayurvedic, Siddha and Unani Tibb system of medicine, to be nominated by the Central Government.

3) The Central Government shall appoint a member of the Board as its Chairman.
4) The nominated members of the Board shall hold office for three years, but shall be eligible for renomination.
5) The Board may, subject to the previous approval of the Central Government, make byelaws fixing a quorum and regulating its own procedure and conduct of all business to be transacted by it.
6) The functions of the Board may be exercised notwithstanding any vacancy therein.
7) The Central Government shall appoint a person to be Secretary of the Board and shall provide the Board with such clerical and other staff as the Central Government considers necessary.

33D. Ayurvedic, Siddha and Unani Drugs Consultative Committee.—

1) The Central Government may constitute an Advisory Committee to be called the Ayurvedic, Siddha and Unani Drugs Consultative Committee to advise the Central Government, the State Governments and the Ayurvedic, Siddha and Unani Drugs Technical Advisory Board on any matter for the purpose of securing uniformity throughout India in the administration of this Act in so far as it relates to Ayurvedic, Siddha or Unani drugs.
2) The Ayurvedic, Siddha and Unani Drugs Consultative Committee shall consist of two persons to be nominated by the Central Government as representatives of that Government and not more than one representative of each State to be nominated by the State Government concerned.
3) The Ayurvedic, Siddha and Unani Drugs Consultative Committee shall meet when required to do so by the Central Government and shall regulate its own procedure.

33E. Misbranded drugs.—For the purposes of this Chapter, an Ayurvedic, Siddha or Unani drugs shall be deemed to be misbranded—

a) If it is so coloured, coated, powered or polished that damage is concealed, or if it is made to appear of better or greater therapeutic value than it really is; or
b) If it is not labelled in the prescribed manner; or
c) If its label or container or anything accompanying the drug bears any statement, design or device which makes any false claim for the drug or which is false or misleading in any particular.

33EE. Adulterated drugs.—For the purposes of this Chapter, an ASU drug shall be deemed to be adulterated—

a) If it consists, in whole or in part, of any filthy, putrid or decomposed substance; or
b) if it has been prepared, packed or stored under insanitary conditions whereby it may have been contaminated with filth or whereby it may have been rendered injurious to health; or
c) If its container is composed, in whole or in part, of any poisonous or deleterious substance which may render the contents injurious to health; or
d) If it bears or contains, for purposes of coloring only, a colour other than one which is prescribed; or
e) If it contains any harmful or toxic substance which may render it injurious to health; or
f) If any substance has been mixed therewith so as to reduce its quality or strength.

33EEA. Spurious drugs.—For the purposes of this Chapter, an ASU drug shall be deemed to be spurious—

a) If it is sold, or offered or exhibited for sale, under a name which belongs to another drug; or
b) If it is an imitation of, or is a substitute for, another drug or resembles another drug in a manner likely to deceive, or bears upon it or upon its label or container the name of another drug, unless it is plainly and conspicuously marked so as to reveal its true character and its lack of identity with such other drug; or
c) If the label or container bears the name of an individual or company purporting to be the manufacturer of the drug, which individual or company is fictitious or does not exist; or
d) If it has been substituted wholly or in part with any other drug or substance; or
e) If it purports to be the product of a manufacturer of whom it is not truly a product.

33EEB. Regulation of manufacture for sale of ASU drugs.—No person shall manufacture for sale or for distribution any Ayurvedic, Siddha or Unani drug except in accordance with such standards, if any, as may be prescribed in relation to that drug.

33EEC. Prohibition of manufacture and sale of certain ASU drug.—From such date as the State Government may, by notification in the Official Gazette, specify in this behalf, no person, either by himself or by any other person on his behalf, shall—

a) Manufacture for sale or for distribution—
   i) Any misbranded, adulterated or spurious ASU drugs;

- Any patent or proprietary medicine, unless there is displayed in the prescribed manner on the label or container thereof the true list of all the ingredients contained in it; and
- Any ASU drug in contravention of any of the provisions of this Chapter or any rule made there under;
b) Sell, stock or exhibit or offer for sale or distribute, any ASU drug which has been manufactured in contravention of any of the provisions of this Act, or any rule made there under;
c) Manufacture for sale or for distribution, any ASU, except under, and in accordance with the conditions of, a license issued for such purpose under this Chapter by the prescribed authority:
Provided that nothing in this section applies to Vaidyas and Hakims who manufacture ASU drugs for the use of their own patients:
Provided further that nothing in this section shall apply to the manufacture, subject to the prescribed conditions, of small quantities of any ASU drug for the purpose of examination, test or analysis.

33EED. Power of Central Government to prohibit manufacture, etc., of ASU drugs in public interest.—Without prejudice to any other provision contained in this Chapter, if the Central Government is satisfied on the basis of any evidence or other material available before it that the use of any ASU drug is likely to involve any risk to human beings or animals or that any such drug does not have the therapeutic value claimed or purported to be claimed for it and that in the public interest it is necessary or expedient so to do then, that Government may, by notification in the Official Gazette, prohibit the manufacture, sale or distribution of such drug.

33-I. Penalty for manufacture, sale, etc., of ASU drug in contravention of this Chapter—whoever himself or by any other person on his behalf—

1) Manufactures for sale or for distribution,—
    a) Any ASU drug—
        i) Deemed to be adulterated under section 33EE, or
        ii) Without a valid license as required under clause (c) of section 33EEC,
Shall be punishable with imprisonment for a term which may extend to one year and with fine which shall not be less than two thousand rupees;
    b) Any ASU drug deemed to be spurious under section 33EEA shall be punishable with imprisonment for a term which shall not be less than one year but which may extend to three years and with fine which shall not be less than five thousand rupees:
Provided that the Court may, for any adequate and special reasons to be mentioned in the judgment, impose a sentence of imprisonment for a term of less than one year and of fine of less than five thousand rupees; or
2) Contravenes any other provisions of this Chapter or of section 24 as applied by section 33H or any rule made under this Chapter, shall be punishable with imprisonment for a term which may extend to three months and with fine which shall not be less than five hundred rupees.

3.4. DRUGS AND COSMETIC AMENDMENT ACT AND ASU DRUGS

- **Amendment to Rule 169:** Considering the growing demand for ASU drugs and to increase palatability longevity & stability of ASU drugs, the matter regarding allowing excipients, preservatives, antioxidants, flavoring agents, chelating agents in ASU drugs was taken up and discussed in various forums. On the recommendation of ASU Drug Technical Advisory Board (ASUDTAB), the amendment to Rule 169 for permitting excipients, preservatives, antioxidants, flavoring agents, chelating agents, etc. in ASU medicines were carried out. The Final Notification has been issued in this regard on 23rd October, 2008.

- **Amendment to Rule 170:** Growing popularity and acceptability of ASU drugs globally and adherence to various regulatory provisions have led to the need for categorization of ASU drugs and other traditional medicines in India and their Pre- Clinical safety guidelines, etc. Since there were no existing guidelines on the subject, a technical Committee was constituted with members of the ICMR and Research Councils. As per the suggestions of the Committee and ASUDTAB recommendations, Rule 170 has been amended regarding issuance of guidelines for the evaluation of ASU Drugs and other traditional medicines of India. Draft Notification has been issued on 24th December, 2008.

- **GMP guidelines for manufacturing of Herbo-mineral-metallic compounds:** To establish the authenticity of raw drugs, minerals and metals in processing of validation and quality control parameters, it is ensured that these formulations are processed and prepared in accordance with clinical tests and for which safety measures are complied with in accordance with GMP guidelines for manufacturing of "Rasaushadhies or Rasamarunthukal and Kushtajat (Herbo-mineral-metallic compounds)" used in ASU medicines. The Final Notification has been issued on 4th March, 2009.

- **Amendment to Rule 161(B):** The potency of ASU preparations is lost/reduced after a certain period of time. Hence, to make full use of these preparations and as per textual reference, ASUDTAB has recommended Shelf life/Expiry date for ASU drugs. Shelf life/Expiry date under rule 161(B) has been amended in respect of ASU medicines. The Final Notification has been issued on 15th October, 2009.

- **Books:** The books entitled *Rastantra Sar Va Siddha Prayog Samgraha Part II (Edition 2006), Ayurvedic Pharmacopoeia of India and its part, Siddha Pharmacopoeia of India and its part* have been amended in Schedule I of the Drugs and Cosmetics Act, 1940. (The Final Notification has been issued on 7th January, 2010.)

- **Amendment to Rule 155(B), 156, 156(A), 157 & Form 13A and Form 26E-I:** In response to the demand of ASU drugs manufacturers for increasing the validity period of GMP license and harmonization from date of issuance of GMP and Schedule 'T' license, and in accordance with ASUDTAB recommendations, Amendment in Rule 155(B), 156, 156(A), 157 & Form 13A and Form 26E-I have been carried out. The validity of GMP Certificate has been extended to five years from three years. GMP certificate in Form 26E-(I) and grant or renewal of license in Form 25-D are proposed for simultaneous issuance. Draft Notification has been issued on 7th January, 2010.

- **Revision of Schedule E (I):** Schedule E of Drugs & Cosmetics Rule 1945 contains a list of poisonous substances under the Ayurvedic (including Siddha) and Unani Systems of medicine. In the list, only some parts of the plants are found poisonous whereas the rest of the plant is not poisonous and some of the names were found incorrect. The matter was examined in detail and finally as per recommendations of ASUDTAB, Schedule E (I) has been revised and necessary amendments in the list of plants and names, etc. for Ayurveda, Unani & Siddha poisonous drugs have been carried out. Draft Notification has been issued on 13th April, 2010.
- **Amendment to Rule 158(B):** The Drugs & Cosmetics Act does not define these ASU products which fall under categories Neutraceutical, food supplements and cosmetics, etc. These ASU plants based medicines/products are also marketed in different doses like extracts, etc. There is an urgent need to regulate standards and quality in this area. There is no regulation existing regarding said ASU products under above said category. The matter was debated in different various committees. As per the recommendation of the Ayurveda, Siddha and Unani Drug Technical Board, the Amendment to Rule 158(B) regarding guidelines for issue of license in respect of ASU drugs have been carried out. Draft Notification has been issued on 3rd May, 2010.

3.5. MAGIC REMEDIES ACT (OBJECTIONABLE ADVERTISEMENTS) ACT, 1954

An Act to control the advertisement of drugs in certain cases, to prohibit the advertisement for certain purposes of remedies alleged to possess magic qualities and to provide for matters connected therewith.

The Drugs and Magic Remedies:

- A medicine for the internal or external use of human beings or animals;
- Any substance intended to be used for or in the diagnosis, cure, mitigation, treatment or prevention of disease in human beings or animals;
- Any article, other than food, intended to affect or influence in any way the structure or any organic function of the body of human beings or animals;
- Any article intended for use as a component of, any medicine, substance or article, referred to in sub-clauses (i), (ii) and (iii);

i) The definition of drug under the Act is comprehensive and even the machine which is an article is – covered by the definition. The definition of 'drug' in section 2 is very comprehensive and exhaustive. Unlike the definition of the drug under Drugs and Cosmetics Act, 1940, it brings within its ambit medicines of all systems including Ayurvedic drugs.

ii) 'Magic remedy' includes –
A talisman, mantra, kavacha and any other charm of any kind which is alleged to possess miraculous powers for or in the diagnosis, cure, mitigation, treatment

or prevention of any disease in human beings or animals or for affecting or influencing in any way the structure or any organic function of the body of human beings or animals;

iii) 'Taking any part in the publication of any advertisement' includes –
- The printing of the advertisement;
- After consultation with the Drug Technical Advisory Board constituted under the Drugs and Cosmetics Act, 1940 (23 of 1940) and, if the Central Government considers necessary, with such other persons having special knowledge or practical experience in respect of Ayurvedic or Unani systems of medicines as that Government deems fit.

3.6. DRUG & COSMETIC ACT AMENDMENT ACT IN THE LAST TWO YEARS (GAZETTE NOTIFICATION ISSUED UNDER THE DRUGS & COSMETICS RULE 1945)

S. No.	G.S.R. No.	Year & Date	Subject
1.	755(E)	23.10.2008	The amendment to Rule 169 for permitting excipients, preservatives, antioxidants, flavoring agents, chelating agents, etc. in Ayurvedic, Siddha and Unani medicines was carried out.
2.	893(E)	24.12.2008	Rule 170 has been amended regarding issuance of guidelines for the evaluation of Ayurvedic, Siddha & Unani Drugs and other traditional medicines of India.
3.	157(E)	04.03.2009	GMP guidelines for manufacturing of "Rasaushadhies or Rasamarunthukal and Kushtajat (Herbo-mineral-metallic compounds)" used in Ayurveda, Siddha and Unani medicines.
4.	764(E)	15.10.2009	Shelf life/Expiry date under rule 161(B) has been amended in respect of Ayurveda, Siddha & Unani medicines.
5.	765(E)	16.10.2009	On manufacturing records of raw materials used by licensed manufacturing units of ASU drugs.
6.	17(E)	07.01.2010	The validity of GMP Certificate has been extended to five years from 3 years.
7.	322(E)	13.04.2010	Schedule E (I) has been revised and necessary amendments in the list of plants and names, etc. for Ayurveda, Unani & Siddha poisonous drugs have been carried out.

CHAPTER 4

AYURVEDIC DRUG MANUFACTURING LICENSE

4.1. LEGAL NECESSITY

As per the Drugs and Cosmetic Act, it is necessary to obtain drug manufacturing license for sale of Ayurvedic Medicines and/or cosmetics in India and for export purposes as well. This is equally applicable across India. In the Drug and Cosmetic Act, in addition to a new license, there is a provision of renewal of license, definition regarding manufacturing and guidelines for Good Manufacturing Practice (GMP).

4.2. THE LICENSING AUTHORITIES

> State Government appoints Licensing Authorities and in such areas as may be specified in this behalf by notification in the Official Gazette (Part XVI rule 152).
> In the states, Director, Ayurveda/AYUSH is the Licensing authority.
> Application can be filed through District Ayurvedic Officer or direct to the Drugs cell of the directorate as the case may be.
> Some of the procedure differs from state to state.

4.3. MINIMUM REQUIREMENTS

> Application on Form 24-D (three copies).
> Blueprints as per specification of rule 157 (dedicated to GMP).
> List of technical staff with qualifications and registration required (BAMS or B.Sc. with two years experience in a reputed registered manufacturing company is the eligibility criteria for manufacturing chemist/technical staff).
> List of medicines for which license is applied with the name and reference book in case of classical drugs.
> Full detailed ingredient, quantity, uses and method of preparation of proprietary medicine. Each product must have separate sheet.
> List of machinery and equipments including laboratory apparatus and equipments.
> Fee-a sum of Rs. 1000 (may vary from state to state).

> Treasury chalan in original with two photocopies (proof of required fee). Fees will be deposited in a particular code assigned for this purpose by the state government and code duly verified the competent authority.

4.4. DOCUMENTS

> *Ownership* – if proprietary, name and address of the owner.
> *Partnership* – partnership deed, power of attorney declaring the name of the signatory.
> *Premises* – Rent agreement if it is rented.
> *Technical officer* – Appointment letter showing full time nature of the technical staff.
> *Consent letter* – agrees to work.
> *Other* – Photo copies of Degree and registration.

4.5. AFFIDAVITS

> **Owner/Proprietor-Affidavit 1**
  * The address.
  * Use of premise for the purpose of manufacturing of drugs.
  * That the premise is not for residential purpose.
  * About raw drugs testing.
  * Finished drugs testing.
  * Maintenance of testing record of raw & finished good.
  * Following all norms of Schedule "T".
  * The owner and partner non-conviction.

> **Owner/Proprietor-Affidavit 2**
  * Manufacturing only approved products.
  * No resemblance with other products regarding their names, formulations, packing and labelling.

> **Technical Officer**
  * Non resemblance with other company's product with the best of his knowledge.
  * Preparation of drugs under direct supervision.
  * Bound to follow the rules and regulations as laid down by the licensing authority.

4.6. TECHNICAL OFFICER'S QUALIFICATION

As per drug act

  * Bachelor of Ayurvedic Medicine and Surgery-B.A.M.S. recognised by the Central Council of Indian Medicine (C.C.I.M.).
  * B.Sc. with two years of Ayurvedic drug manufacturing experience from any reputed university.

4.7. MINIMUM SPACE

> Office space 100 sq feet
> Raw drug store 150 sq feet
> Finished medicine store 150 sq feet
> Laboratory 150 sq feet
> Rejected drug store 100 sq feet
> Churna room 200 sq feet
> Furnance 200 sq feet
> Packaging area-
>   o Packaging and other space should be as per *kaplana* to use during manufacturing.

4.8. DRUG APPROVAL

>*Classical Ayurvedic drugs*
  * Based on the classical text of Ayurveda mentioned in first schedule of the Drugs & Cosmetic Act.

>*Proprietary Ayurvedic drugs*
  * Based upon personal experience and/or research.

4.9. IMPORTANT NOTICE FOR ASU DRUG MANUFACTURES

Rule 157A of Drug and Cosmetics Rules, 1945 provides that each licensed manufacturing unit of the Ayurveda, Siddha and Unani Drugs shall keep a record of raw materials used by it in the proforma given for the purpose in respect of all raw materials utilized by it in the preceding financial year and shall submit the same by the 30th day of June of the succeeding financial year to the State Drug Licensing Authority of Ayurveda, Siddha and Unani drugs and to the NMPB or an agency nominated by the NMPB for this purpose.

NMPB nominated M/s Centre for Research, Planning and Action (CERPA), 16 Dakshineshwar, 10-Hailey Road, New Delhi as the agency for the purpose of collection of the said proforma with details for the years 2010–11, 2011–12 and 2012–13.

4.10. AYUSH NOTIFICATION NO.: G.S.R. 512(E) (09-JUL-08) DRUG AND COSMETICS (FIRST AMENDMENT) RULES, 2008

1. Whereas the draft of certain rules to further amend the Drugs and Cosmetics Rules, 1945 was published as required by section 33N of the Drugs and Cosmetics Act, 1940 (23 of 1940), in the Gazette of India, Extraordinary, dated the 19th October, 2006, vide Number GSR 651(E) inviting objections and suggestions from persons likely to be affected thereby and notice was given that the said draft will be taken into consideration after the expiry of a period of forty-five days from the date on which copies of the Official Gazette containing the said notification were made available to the public;

And whereas, the said Gazette was made available to the public on the 18th October, 2006;

And whereas, objections and suggestions received from the public on the said draft rules Drugs and Cosmetics Act, 1940 (23 of 1940) have been considered by the Central Government;

Now, therefore, in exercise of the powers conferred by section 33-N of the Drugs and Cosmetics Act, 1940 (23 of 1940) the Central Government, after consultation with the Ayurveda, Siddha and Unani Drugs Technical Advisory Board, hereby makes the following rules further to amend the Drugs and Cosmetics Rules, 1945, namely:

1) These rules may be called the Drug and Cosmetics (First Amendment) Rules, 2008.
2) They shall come into force from the date of their publication in the Official Gazette.

2. In the Drug and Cosmetics Rules, 1945 (herein referred to as the said rules), after rule 157, the following rule shall be inserted, namely:-

"157A. Maintaining of records of raw material used by a licensed manufacturing unit of Ayurveda, Siddha and Unani drugs during the preceding financial year. Each licensed manufacturing unit of Ayurveda or Siddha or Unani drugs shall keep a record of raw material used by each licensed manufacturing unit of Ayurveda, Siddha or Unani drugs, in the performa given in Schedule TA in respect of all raw materials utilized by that unit in the manufacture of Ayurveda or Siddha or Unani drugs in the preceding financial year, and shall submit the same by the 30th day of June of the succeeding financial year to the State Drug Licensing Authority of Ayurveda, Siddha and Unani drugs and to the National Medicinal Plants Board or any agency nominated by the National Medicinal Plant Board for this purpose."

3. In the said rules, after Schedule T, following Schedule shall be inserted.

CHAPTER 5

GOOD MANUFACTURING PRACTICES FOR ASU MEDICINES (SCHEDULE-T)

5.1. THE GOOD MANUFACTURING PRACTICES (GMP)

The Good Manufacturing Practices (GMP) are prescribed as follows in Part I and Part II to ensure that:

- Raw materials used in the manufacture of drugs are authentic, of prescribed quality and are free from contamination.
- The manufacturing process is as has been prescribed to maintain the standards.
- Adequate quality control measures are adopted.
- The manufactured drug which is released for sale is of acceptable quality. (v)
- To achieve the objectives listed above, each licensee shall evolve methodology and procedures for following the prescribed process of manufacture of drugs which should be documented as a manual and kept for reference and inspection. However, under IMCC Act 1970, registered Vaidyas, Siddhas and Hakeems who prepare medicines on their own to dispense to their patients and are not selling such drugs in the market are exempted from the purview of G.M.P.

5.2. GOOD MANUFACTURING PRACTICES (PART-1)

Factory Premises

The manufacturing plant should have adequate space for:-

i) Receiving and storing raw material
ii) Manufacturing process areas
iii) Quality control section
iv) Finished goods store
v) Office
vi) Rejected goods/drugs store

General Requirements

1.1(A) Location and surroundings – The factory building for the manufacture of Ayurveda, Siddha and Unani medicines shall be so situated and shall have such construction as to avoid contamination from open sewerage, drain, public lavatory or any factory which produces disagreeable or obnoxious odour or fumes or excessive soot, dust or smoke.

1.1(B) Buildings – The building used for factory shall be such as to permit production of drugs under hygienic conditions and should be free of cobwebs and insects/rodents. It should have adequate provision of light and ventilation. The floor and the walls should not be damp or moist. The premises used for manufacturing, processing, packaging and labelling will be in conformity with the provisions of the Factory Act. It shall be located so as to be:

i) Compatible with other manufacturing operations that may be carried out in the same or adjacent premises.
ii) Adequately provided with working space to allow orderly and logical placement of equipment and materials to avoid the risk of mix-up between different drugs or components thereof and control the possibility of cross contamination with other drugs or substances and avoid the risk of omission of any manufacturing or control step.
iii) Designed, constructed and maintained to prevent entry of insects and rodents. Interior surface (walls, floors and ceilings) shall be smooth and free from cracks and permit easy cleaning and disinfection. The walls of the room in which the manufacturing operations are carried out should not allow light or water to enter or pass through. The flooring shall be smooth and even and shall be such as not to permit retention or accumulation of dust or waste products.
iv) Provided with proper drainage system in the processing area. The sanitary fittings and electrical fixtures in the manufacturing area shall be proper and safe.
v) Furnace/Bhatti section could be covered with tin roof and proper ventilation, but sufficient care should be taken to prevent flies and dust.
vi) There should be fire safety measures and proper exits should be there.
vii) Drying space – There should be a separate space for drying of raw material, in process medicine or medicines which require drying before packing. This space will be protected from flies/insects/dusts, etc., by proper flooring, wire mesh windows, glass prances or other material.

1.1(C) **Water Supply** – The water used in manufacture shall be pure and of potable quality. Adequate provision of water for washing the premises shall be made.

1.1(D) **Disposal of Waste** – From the manufacturing sections and laboratories the waste water and the residues which might be prejudicial to the workers or public health shall be disposed off after suitable treatment as per guidelines of pollution control authorities to render them harmless.

1.1(E) **Containers' Cleaning** – In factories where operations involving the use of containers such as glass bottles, vials and jars are conducted, there shall be adequate arrangement separated from the manufacturing operations for the washing, cleaning and drying of such containers.

1.1(F) **Stores** – Storage should have proper ventilation and shall be free from dampness. It should provide independent, adequate space for storage of different types of material, such as raw material, packaging material and finished products.

1.1(F)(A) **Raw Materials** – All raw materials procured for manufacturing will be stored in the raw materials store. The manufacture based on the experience and the characteristics of the particular raw material used in Ayurveda, Siddha and Unani system shall decide the use of appropriate containers which would protect the quality of the raw material as well as prevent it from damage due to dampness, microbiological contamination or rodent and insect infestation, etc.

If certain raw materials require such controlled environmental conditions, the raw materials stores may be sub-divided with proper enclosures to provide such conditions by suitable cabinization. While designing such containers, cabins or areas in the raw materials store, care may be taken to handle the following different categories of raw materials:

1) Raw material of metallic origin.
2) Raw material of mineral origin.
3) Raw material from animal source.
4) Fresh Herbs.
5) Dry Herbs or plant parts.
6) Excipients, etc.
7) Volatile oils/perfumes & flavours.
8) Plant concentrates/extracts and exudates/resins.

Each container used for raw material storage shall be properly identified with the label which indicates the name of the raw material, source of supply and will also clearly state the status of raw material such as 'UNDER TEST' or 'APPROVED' or 'REJECTED'. The labels shall further indicate the identity of the particular supply in the form of Batch No. or Lot No. and the date of receipt of consignment.

All the raw materials shall be sampled and tested either by the in-house Ayurvedic, Siddha and Unani experts (Quality control technical person) or by the laboratories approved by the Government and shall be used only on approval after verifying. The rejected raw material should be removed from the other raw materials store and should be kept in a separate room. Procedure of 'First in, first out' should be adopted for raw materials wherever necessary. Records of the receipt, testing and approval or rejection and use of raw material shall be maintained.

1.1(F)(B) **Packaging Materials** – All packaging materials such as bottles, jars, capsules, etc. shall be stored properly. All containers and closures shall be adequately cleaned and dried before packing the products.

1.1(F)(C) **Finished Goods Stores** – The finished goods transferred from the production area after proper packaging shall be stored in the finished goods stores within an area marked "Quarantine". After the quality control laboratory and the experts have checked the correctness of finished goods with reference to its packing/labelling as well as the finished product quality as prescribed, then it will be moved to "Approved Finished Goods Stock" area. Only approved finished goods shall be dispatched as per marketing requirements. Distribution records shall be maintained as required. If any Ayurvedic, Siddha and Unani drug needs special storage conditions, finished goods store shall provide necessary environmental requirements.

1.1(G) **Working Space** – The manufacturing area shall provide adequate space (manufacture and quality control) for orderly placement of equipment and material used in any of the operations for which these are employed so as to facilitate easy and safe working and to minimize or to eliminate any risk of mix-up between different drugs, raw materials and to prevent the possibility of cross-contamination of one drug by another drug that is manufactured, stored or handled in the same premises.

1.1(H) **Health, Clothing, Sanitation and Hygiene of Workers** – All workers employed in the Factory shall be free from contagious diseases. The clothing of the workers shall consist of proper uniform suitable to the nature of work and the climate and shall be clean. The uniform shall also include a cloth or synthetic covering for hands, feet and head wherever required. Adequate facilities for personal cleanliness such as clean towels, soap and scrubbing brushes shall be provided. Separate provision shall be made for lavatories to be used by men and women, and such lavatories shall be located at places separated from the processing rooms. Workers will also be provided facilities for changing their clothes and to keep their personal belongings.

1.1(I) **Medical Services** – The manufacturer shall also provide:- (a) Adequate facilities for first aid; (b) Medical examination of workers at the time of employment and periodical check up thereafter by a physician once a year, with particular attention being devoted to freedom from infections. The records thereof shall be maintained.

1.1(J) **Machinery and Equipments** – For carrying out manufacturing depending on the size of the operation and the nature of product manufactured, suitable equipment either manually operated or operated semi-automatically (electrical or team based) or fully automatic machinery shall be made available. These may include machines for use in the process of manufacture such as crushing, grinding, powdering, boiling, mashing, burning, roasting, filtering, drying, filling, labelling and packing, etc.

To ensure ease of movement of workers and orderliness in operations a suitably adequate space will be ensured between two machines or rows of machines. These machinery and equipments recommended is indicated in Part II-A. Proper standard operational procedures (SOPs) for cleaning, maintaining and performance of every machine should be laid down.

1.1(K) **Batch Manufacturing Records** – The licensee shall maintain the batch manufacturing record of each batch of Ayurvedic, Siddha and Unani drugs manufactured irrespective of the type of product manufactured (classical preparation or patent and proprietary medicines). Manufacturing records are required to provide and account of the list of raw materials and their quantities obtained from the store, tests conducted during the various stages of manufacture like taste, colour, physical characteristics and chemical tests as may be necessary or indicated in the approved books of Ayurveda, Siddha and Unani mentioned in the First Schedule of the Drugs and Cosmetics Act, 1940 (23 of 1940). These tests may include any in-house or pharmacopoeial test adopted by the manufacturer in the raw material or in the process material and in the finished product. These records shall be duly signed by Production and Quality Control Personnel respectively.

Details of transfer of manufactured drugs in the finished products store, including dates and quantity of drugs transferred along with record of testing of the finished product, if any, and packaging, records shall be maintained. Only after the manufactured drugs have been verified and are of accepted quality shall they be allowed to be cleared for sale. It should be essential to maintain the record of date, manpower, machine and equipments used and to keep in the process the record of various *shodhana* (purification), *bhavana* (trituration), burning in fire and specific grindings in terms of internal use.

1.1(L) **Distribution Records** – Records of sale and distribution of each batch of Ayurveda, Siddha and Unani Drugs shall be maintained in order to facilitate prompt and complete recall of the batch, if necessary. The duration of record keeping should be the date of expiry of the batch. Certain categories of Ayurvedic, Siddha and Unani medicines like Bhasma, Rasa, Kupipakva, Parpati, Sindura, Karpu/Uppu/Puram, Kushta, Asava-arista, etc. do not have an expiry date; in contrast, their efficacy increases with the passage of time. Hence, records need to be maintained up to 5 years of the exhausting of stock.

1.1(M) **Record of Market Complaints** – Manufacturers shall maintain a register to record all reports of market complaints received regarding the products sold in the market. The manufacturer shall enter all data received on such market complaints, investigations carried out by the manufacturers regarding the complaint as well as any corrective action initiated to prevent recurrence of such market complaints shall also be recorded. Once in a period of six months the manufacturer shall submit the record of such complaints to the Licensing Authority. The register maintained by the pharmacy shall also be available for inspection during any inspection of the premises. Reports of any adverse reaction resulting from the use of Ayurvedic, Siddha and Unani drugs shall also be maintained in a separate register by each manufacturer. The manufacturer shall investigate the adverse reaction to find if the same is due to any defect in the product, and whether such a reaction is already reported in the literature or if it is a new observation.

1.1(N) **Quality Control** – Every licensee is required to provide a facility for quality control section in his own premises or through a Government-approved testing laboratory. The test shall be as per the Ayurveda, Siddha and Unani pharmacopoeial standard. Where the tests are not available, the test should be performed according to the manufacturer's specification or other information available. The quality control section shall verify all the raw materials, monitor in process, quality checks and control the quality of finished product being released to finished goods store/warehouse.

Preferably for such quality control there should preferably be a separate expert. The quality control section shall have the following facilities:

1) There should be a 150 sq feet area for the quality control section.
2) For identification of raw drugs, reference books and reference samples should be maintained.
3) Manufacturing record should be maintained for the various processes.
4) To verify the finished products, controlled samples of finished products of each batch will be kept till the expiry date of the product.
5) To supervise and monitor the adequacy of conditions under which raw materials, semi-finished products and finished products are stored.
6) Keep records in establishing shelf life and storage requirements for the drugs.
7) Manufacturers who are manufacturing patent proprietary Ayurveda, Siddha and Unani medicines shall provide their own specification and control references in respect of such formulated drugs.
8) The record of a specific method and procedure of preparation, that is, "Bhavana", "Mardana" and "Puta" and the record of every process carried out by the manufacturer shall be maintained.
9) The standards for identity, purity and strength as given in respective pharmacopoeias of Ayurveda, Siddha and Unani systems of medicines published by Government of India shall be complied with.
10) All raw materials will be monitored for fungal and/or bacterial contamination with a view to minimize such contamination.
11) Quality control section will have a minimum of-
    i) One person with Ayurveda/Unani/Siddha qualification recognized under Schedule II of Indian Medicine Central Council Act, 1970. Two other persons, one each with a Bachelor qualification in Botany/Chemistry/Pharmacy could be on part-time or on a contractual basis.
    ii) The manufacturing unit shall have a quality control section as explained under section 35. Alternatively, these quality control provisions will be met by getting tested from a recognized laboratory for Ayurveda, Siddha and Unani drugs; under Rule 160-A of the Drugs and Cosmetics Act. The manufacturing company will maintain all the records of various tests done from an outside recognized laboratory.
    iii) List of equipment recommended is indicated in Part II-C.

Requirement for Sterile Product

A) **Manufacturing Areas** – For the manufacture of sterile Ayurvedic, Unani and Siddha drugs, separate enclosed areas specifically designed for the purpose shall be provided. These areas shall be provided with air locks for entry and shall be essentially dust free and ventilated with an air supply. For all areas where aseptic manufacture has to be carried out, the air supply shall be filtered through bacteria retaining filters (HEPA Filters) and shall be at a pressure higher than the adjacent areas. The filters shall be checked for performance on installation and periodically thereafter the record of checks shall be maintained. All the surfaces in sterile manufacturing areas shall be designed to facilitate cleaning and disinfection. For the sterile manufacturing routine, microbial counts of all Ayurvedic, Siddha and Unani drug manufacturing areas shall be carried out during operations. The results of such count shall be checked against established in-house standards and record maintained.

Access to manufacturing areas shall be restricted to the minimum number of authorized personnel. The special procedure to be followed for entering and leaving the manufacturing areas shall be written down and displayed. For the manufacturing of Ayurvedic, Siddha and Unani drug that can be sterilized in their final containers, the design of the areas shall preclude the possibility of the products intended for sterilization being mixed with or taken to be products already sterilized. In case of terminally sterilized products, the design of the areas shall preclude the possibility of mix-up between non-sterile products.

B) **Precautions against contamination and mixing:**
   a) Carrying out manufacturing operations in a separate block of an adequately isolated building or operating in an isolated enclosure within the building.
   b) Using appropriate pressure differential in the process area.
   c) Providing a suitable exhaust system.
   d) Designing laminar flow sterile air system for sterile products.
   e) The germicidal efficiency of UV lamps shall be checked and recorded, indicating the burning hours or checked - intensity.
   f) Individual containers of liquids and ophthalmic solutions shall be examined against black-white background fitted with diffused light after filling to ensure freedom from contamination with foreign suspended matter.
   g) Expert technical staff approved by the Licensing Authority shall check and compare the actual yield against theoretical yield before final distribution of the batch. All process controls as required under master formula including room temperature, relative humidity, volume filled, leakage and clarity shall be checked and recorded.

5.3. PART-II

A. LIST OF RECOMMENDED MACHINERY, EQUIPMENT AND MINIMUM MANUFACTURING PREMISES REQUIRED FOR THE MANUFACTURE OF VARIOUS CATEGORIES OF AYURVEDIC, SIDDHA SYSTEM OF MEDICINES

One machine indicated for one category of medicine could be used for the manufacturing of another category of medicine also. Similarly, some of the manufacturing areas like powdering, furnace, packing of liquids and Avaleha, Paks, could also be shared for these items.

Sl. No.	Category of Medicine	Minimum Manufacturing Space Required*	Machinery/equipment Recommended
1.	Anjana/Pisti	100 sq feet	Karel/mechanized/motorized, karel. End runner/ Ball-Mill Sieves/Shifter.
2.	Churna/Nasya/ Manjan/Lepa/Kwath Churn	200 sq feet	Grinder/disintegrator/Pulveriser/Powder mixer/ sieves/shifter.
3.	Pills/Vati/Gutika Matirai and tablets	100 sq feet	Ball Mill, Mass mixer/powder mixer, Granulator, drier, tablet compressing machine, pill/vati cutting machine, stainless steel trays/container for storage and sugar coating, polishing pan in case of sugar-coated tablets, mechanised chattoo (for mixing guggulu) where required.
4.	Kupi pakava/ Ksara/Parpati/ LavanaBhasma Satva/ Sindura Karpu/Uppu/ Param	150 sq feet	Bhatti, Karahi/Stainless steel Vessels/Patila Flask, Multani Matti/Plaster of Paris, Copper Rod, Earthern container, Gaj Put Bhatti, Muffle furnace (Electrically operated) End/Edge Runner, Exhaust Fan, Wooden/S.S. Spatula.
5.	Kajal	100 sq feet	Earthern lamps for collection of Kajal, Triple Roller Mill, End Runner, Sieves, S.S. Patila, Filling/packing and manufacturing room should be provided with exhaust fan and ultra violet lamps.
6.	Capsules	100 sq feet	Air Conditioner, De-humidifier, hygrometer, thermometer, Capsule filling machine and chemical balance.
7.	Ointment/Marham Pasai 1	100 sq feet	Tube filling machine, Crimping Machine/Ointment Mixer, End Runner/Mill (Where required) S.S. Storage Container S.S. Patila.
8.	Pak/Avaleh/Khand/ Modak/Lakayam	100 sq feet	Bhatti section fitted with exhaust fan and should be fly proof, Iron Kadahi/S.S. Patila and S.S. Storage container.
9.	Panak, Syrup/Pravahi Kwath Manapaku	150 sq feet	Tincture press, exhaust fan fitted and fly proof, Bhatti section, Bottle washing machine, filter press/ Gravity filter, liquid filling machine, P.P. Capping Machine.
10.	Asava/Arishta	200 sq feet	Same as mentioned above. Fermentation tanks, containers and distillation plant where necessary, Filter Press.
11.	Sura	100 sq feet	Same as mentioned above plus Distillation plant and Transfer pump.

Contd....

Contd.

12.	Ark Tinir	100 sq feet	Maceration tank, Distillation plant, Liquid filling tank with tap/Gravity filter/Filter press, Visual inspection box.
13.	Tail/Ghrit Ney	100 sq feet	Bhatti, Kadahi/S.S. Patila S.S. Storage Containers, Filtration equipment, filling tank with tap/Liquid filling machine.
14.	Aschyotan/Netra Malham Panir/Karn Bindu/Nasabindu	100 sq feet	Hot air oven electrically heated with thermostatic control, kettle gas or electrically heated with suitable mixing arrangements, collation mill, or ointment mill, tube filling.

*1200 Square feet covered area with separate cabins or partitions for each activity. If Unani medicines are manufactured in same premises an additional area of 400 sq feet will be required.

Important Note: Each manufacturing unit will have a separate area for Bhatti, furnace boilers, puta, etc. This will have proper ventilation, removal of smoke, prevention of flies, insects, dust, etc. The furnace section could have a tin roof.

B. LIST OF MACHINERY, EQUIPMENT AND MINIMUM MANUFACTURING PREMISES REQUIRED FOR THE MANUFACTURE OF VARIOUS CATEGORIES OF UNANI SYSTEM OF MEDICINES

C. LIST OF EQUIPMENT RECOMMENDED FOR IN-HOUSE QUALITY CONTROL SECTION (Alternatively, the unit can get testing done from the Government approved laboratory)

A. CHEMISTRY SECTION

1. Alcohol Determination Apparatus (complete set).
2. Volatile Oil Determination Apparatus.
3. Boiling Point Determination Apparatus.
4. Melting Point Determination Apparatus.
5. Refractometer.
6. Polarimeter.
7. Viscometer.
8. Tablet Disintegration Apparatus.
9. Moisture Meter.
10. Muffle Furnace.
11. Electronic Balance.
12. Magnetic Stirrer.
13. Hot Air Oven.
14. Refrigerator.
15. Glass/Steel Distillation Apparatus.
16. LPG Gas Cylinders with Burners.
17. Water Bath (Temperature controlled).

18. Heating Mantles/Hot Plates.
19. TLC Apparatus with all accessories (Manual).
20. Paper Chromatography apparatus with accessories.
21. Sieve size 10 to 120 with Sieve shaker.
22. Centrifuge Machine.
23. Dehumidifier.
24. pH Meter.
25. Limit Test Apparatus.

B. PHARMACOGNOSY SECTION
1. Microscope Binocular.
2. Dissecting Microscope.
3. Microtome.
4. Physical Balance.
5. Aluminium Slide Trays.
6. Stage Micrometer.
7. Camera Lucida (Prism and Mirror Type).
8. Chemicals, Glassware, etc.

Important Note: The above requirements of machinery, equipments, space is made subject to the modification at the discretion of the Licensing Authority; if the Licensing Authority is of the opinion that, having regard to the nature and extent of the manufacturing operations it is necessary to relax or alter them in the circumstances in a particular case.

5.4. STATUS OF GMP COMPLIANCE OF AYURVEDA SIDDHA UNANI MANUFACTURING UNITS

S. No.	State	Total No. of Units	GMP Complying Units	Non-GMP Complying Units	Legal Notices Issued/ Cancelled
1.	Andhra Pradesh	593	543	50	-
2.	Assam	62	04	58	01
3.	Chhatisgarh	31	31	29	03
4.	Dadra & Nagar Haveli	05	02	03	-
5.	Delhi	110	110	-	-
6.	Goa	9	07	02	02
7.	Gujarat	460	460	00	00
8.	Haryana	254	182	72	06
9.	Himachal Pradesh	151	145	03	04
10.	Jammu & Kashmir	16	06	10	-

Contd....

Contd.

11.	Karnataka	167	157	10	25
12.	Kerala	1271	510	761	51
13.	Madhya Pradesh	672	343	321	08
14.	Maharashtra	675	275	400	-
15.	Manipur	05	-	-	-
16.	Orissa	329	92	56	44
17.	Punjab	270	256	14	15
18.	Rajasthan	285	285	-	-
19.	Sikkim	03	01	01	01
20.	Tamil Nadu	668	116	552	21
21.	Uttar Pradesh	3683	1706	1977	1659
22.	Uttarakhand	156	38	118	17
23.	West Bengal	213	133	80	80
	Total	**10088**	**5402**	**4517**	**2468**

Source: http://indianmedicine.nic.in/showfile.asp?lid=315.

CHAPTER 6
AYURVEDIC DRUG INDUSTRY

6.1. INTRODUCTION

India has perhaps the world's oldest as well as largest tradition of systems of medicine. The term Indian Systems of Medicine covers both the systems which originated in India as well as outside, but got adopted in India in the course of time. These systems are Ayurveda, Siddha, Unani, Homoeopathy, Yoga, and Naturopathy. They have become a part of the culture and traditions of India. India with its strong base in traditional knowledge of herbal medicine and vast plant biodiversity has a great potential in this sector.

The codified Indian system of medicine puts to use raw drugs obtained from around 2,400 plant species. The number of raw drugs in trade is still less, i.e., 1289 botanicals obtained from 960 plant taxa. The highest proportion of the traded medicinal plant species is used under the Ayurvedic system.

6.2. CURRENT SCENERIO OF AYURVEDIC DRUG INDUSTRY

The recent upsurge in the use of Ayurvedic medicines has led to a sudden increase in Ayurvedic manufacturing units. In India, there are about 14 well-recognized and 86 medium scale manufacturers of Ayurvedic drugs. Other than this about 8,000 licensed small manufacturers in India are on record. In addition, thousands of Vaidyas (Ayurvedic physicians) also have their own miniature manufacturing facilities. The estimated current annual production of herbal drugs is around Rs. 3500 crores. This section gives an overview of the rapidly growing Indian herbal industry followed by the legal parameters encompassing the manufacturing of herbal drugs.

The turnover of AYUSH industry is estimated to be more than Rs. 8800 crore. The domestic market of Indian Systems of Medicine & Homoeopathy (ISM&H) is of the order of Rs. 4000 Crore with a total consumption of all botanicals to a figure of 177000 MT, which is expanding day by day. The total annual turnover of the Ayurvedic drug manufacturing industry is estimated to be around Rs. 3,500 Crore. Besides this, there is also a growing demand for natural products including items of medicinal value/pharmaceuticals, food supplements and cosmetics in both domestic and international markets. India with its diversified biodiversity has a tremendous potential and advantage in this emerging area.

6.3. MANUFACTURING UNITS

There is a complex of a large number of manufacturing units using herbal material for various purposes. Whereas the largest number of such manufacturing units are registered as 'pharmaceuticals', there are others that are engaged in making plant based cosmetics and food supplements. Even within the pharmaceutical units, there are manufacturers of Ayurveda, Siddha, Unani and Homeopathic formulations (Fig. 1) with a few even making western medicines. Another group of manufacturing units is engaged in making extracts and distilling oils for use by other industries and for exports. Raw materials for all these diverse industries are largely derived from wild sources.

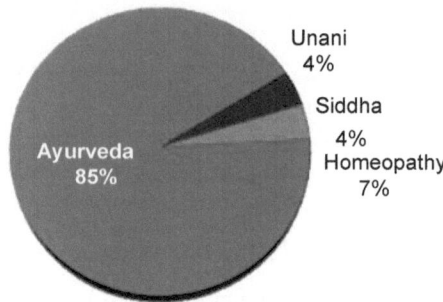

Fig. 1. Proportion of manufacturing units of ASU & H.
Source: Dept. of AYUSH, 2007.

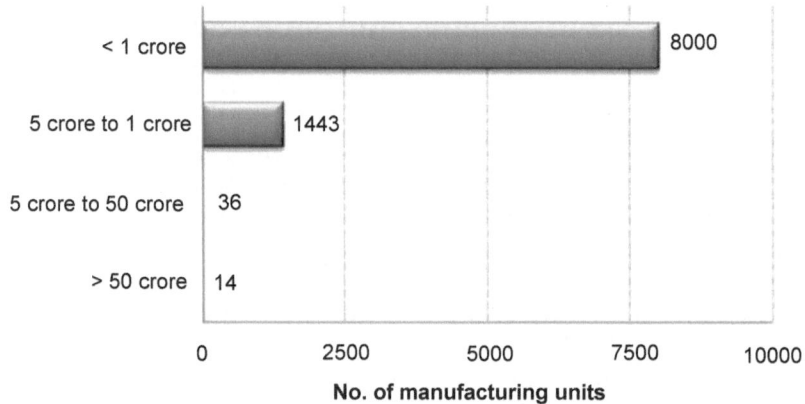

Fig. 2. Herbal Manufacturing Units.
Source: National Medicinal Plant Board, 2008.

6.4. LICENSED PHARMACIES UNDER AYUSH IN 2013

As of 1 April 2013, there were 8896 AYUSH drug manufacturing units (licensed pharmacies) in the country. Out of these, 99.5% of the licensed pharmacies were controlled by nongovernment bodies, and only 0.5% licensed pharmacies were in the government sector. System-wise distribution of these units were quite uneven as 87% licensed pharmacies belonged to Ayurveda, whereas 5.5%, 3.9% and 3.6% were under Unani, Siddha and Homoeopathy systems respectively.

There were 26 states and union territories of the country which have AYUSH licensed pharmacies as of 1 April 2013. No AYUSH drug manufacturing unit has been reported in the states of Jharkhand, Manipur, Meghalaya, Mizoram, Nagaland, Tripura, and in the Union Territories of Andaman & Nicobar Islands, Dadar & Nagar Haveli and Lakshadweep. Uttar Pradesh has the maximum number (2338) of AYUSH licensed pharmacies. The states of Andhra Pradesh, Gujarat, Kerala, Maharashtra, Madhya Pradesh, Tamil Nadu and Uttar Pradesh each were having more than 5% of AYUSH licensed pharmacies. Maximum number of Ayurveda and Unani pharmacies, viz., 2033 and 269 respectively, exist in Uttar Pradesh. Licensed pharmacies under Siddha systems exist in the states of Tamil Nadu (332), Puducherry (10) and Kerala (2). A majority of Homoeopathy licensed pharmacies 25% exist in the state of West Bengal followed by Bihar (14.9%), Maharashtra (12%), Uttar Pradesh (11%), Kerala (7.4%) and Andhra Pradesh (6.8%).

The States/UTs of Jharkhand, Manipur, Meghalaya, Mizoram, Nagaland, Tripura, Andaman & Nicobar Islands, D&N Haveli and Lakshadweep reported to have no licensed pharmacy under any AYUSH system. Besides, there was no Homoeopathy licensed pharmacy in Arunachal Pradesh, Chhattisgarh, Goa, Haryana, Punjab, Rajasthan, Sikkim, Uttarakhand, Chandigarh and Daman & Diu. 158 Unani licensed pharmacies exist in the states of Andhra Pradesh, Bihar, Delhi, Haryana, Jammu & Kashmir, Karnataka, Kerala, Madhya Pradesh, Maharashtra, Tamil Nadu, Uttar Pradesh, Uttarakhand and West Bengal.

Except Arunachal Pradesh, Jharkhand, Manipur, Meghalaya, Mizoram, Nagaland, Tripura, A&N Islands, D&N Haveli and Lakshadweep, the rest of the states have pharmacies compliant with Good Manufacturing Practices (GMP). Out of all drug manufacturing units, GMP-compliant units comprise 78.2% (Non-GM compliant units comprises 21.8%) of the total drug manufacturing units; and within the total GM compliant units, 92.8% were Ayurveda drug manufacturing units and only 3.0%, 1.2% and 3.0% were Unani, Siddha and Homoeopathy drug manufacturing units respectively. The states/UTs having centpercent GMP-compliant drug manufacturing units were Chhattisgarh, Delhi, Goa, Gujarat, Himachal Pradesh, Sikkim, Chandigarh, Daman & Diu and Puducherry. The other states having higher (greater than 75%) proportion of GMP-compliant units were Andhra Pradesh (99.1%), Haryana (93.5%), Karnataka (95.6%), Madhya Pradesh (82.4%), Punjab (97.7%), Rajasthan (98.4%), Uttar Pradesh (89.4%), Uttarakhand (83.3%) and West Bengal (88.50%), Whereas, the only state/union territory having less than 25% GMP-compliance was Tamil Nadu (24.4%). There had been a significant system-wise variation in the proportion of GMP-compliant units, as there were 83.35%, 43.71%, 24.71% and 64.09% GMP-compliant drug manufacturing units under Ayurveda, Unani, Siddha and Homoeopathy systems respectively.

The States/UTs of Andhra Pradesh, Chhattisgarh, Delhi, Goa, Gujarat, Himachal Pradesh, Sikkim, Uttar Pradesh, Chandigarh, Daman & Diu and Puducherry were having cent-percent GM compliant drug manufacturing units under Ayurveda system. Other states which have higher (greater than 75%) proportion of GM compliant units under Ayurveda were Haryana, Karnataka, Madhya Pradesh, Punjab, Rajasthan, Uttarakhand and West Bengal. All Unani drug manufacturing units were GMP-compliant in the states of Andhra Pradesh, Delhi, Haryana, Karnataka, Kerala, Madhya Pradesh, Uttarakhand and West Bengal. All Siddha drug manufacturing units were GMP-compliant with the state of Puducherry only, whereas, 50% and 22.3% were GMP-compliant with the state of Kerala and Tamil Nadu respectively. Likewise, in the states of Delhi, Gujarat, Himachal Pradesh, Maharashtra, Uttar Pradesh and Puducherry, all Homeopathic drug manufacturing units were GMP-compliant.

CHAPTER 7

PHARMACOVIGILANCE OF AYURVEDIC DRUGS

7.1. PHARMACOVIGILANCE (PV or PhV)

Pharmacovigilance, also known as Drug Safety, is the pharmacological science relating to the collection, detection, assessment, monitoring, and prevention of adverse effects from pharmaceutical products. In other words, pharmacovigilance focuses on adverse drug reactions (ADRs), which are defined as any response to a drug which is noxious and unintended, including lack of efficacy.

7.2. ADVERSE DRUG REACTION TO AYURVEDA MEDICINES

Although a technical term equivalent to "pharmacovigilance" does not feature in Ayurvedic texts, the spirit of pharmacovigilance is vibrant throughout Ayurveda's classical literature. The Brihattrayi and Laghutrayi repeatedly emphasize the major goals of pharmacovigilance, to improve patient care and safety during treatment, and thus to promote rational use of medications. These are recurrent themes of Dravyaguna, Rasa Shastra and Bhaishjya Kalpana and Chikitsa. It is probable that these basic principles of Ayurveda gave rise to the common belief that Ayurvedic medicines are safe.

The Ayurvedic literature gives details of drug-drug and drug-diet incompatibilities based on elaborately described qualitative differences in ingredients or quantitative proportions. These factors undoubtedly prevent the onset of many unfortunate reactions. Ayurveda's Anupan therapeutic method and Shodhan pharmaceutics principles probably also contribute to the prevention of many undesired and unforeseen events. Prevention of this kind is a major goal of pharmacovigilance programs.

According to a survey conducted by the National Center for Complementary and Alternative Medicine (NCCAM) in the USA, only about 751,000 people in the United States had ever used Ayurveda and only 154,000 people had used it within the past 12 months. Broadly speaking, two categories of medicines labeled as "Ayurvedic" are available in the market: firstly, classical Ayurvedic formulations, which are as per described in the Ayurveda Samhitas and secondly, patent and proprietary formulations made of extracts of herbs.

Classical Ayurveda prescribes metals and minerals as medicines given as Bhasma or in combination with plants as herbo-mineral formulations (e.g., Aarogyavardhini). Manufacturing procedures for these medicines are stringent, and adverse reactions are described when precautions are not taken while manufacturing and administering these medicines. Although these medicines are widely used in India, doubts about their long-term safety come up due to the presence of toxic metals in them and there are reports related to adverse reactions.

Nearly 21 percent of the Ayurvedic medicines tested were found to contain detectable levels of lead (most common), mercury, or arsenic. All metal-containing products exceeded one or more standards for acceptable daily metal intake. The prevalence of metal-containing products did not differ significantly by country of manufacture. Rasa Shastra products were more than twice as likely as non-Rasa Shastra products to contain metals, and several Rasa Shastra medicines manufactured in India could result in lead and/or mercury ingestion 100 to 10,000 times greater than acceptable limits.

7.3. ASSESSMENT OF ADVERSE REACTIONS TO AYURVEDIC MEDICINES

Although several scales are available for causality assessment, applying them for Ayurvedic medicines and ascribing causality is perhaps the greatest challenge for several reasons, including:

1. Information related to adverse effects is scattered in Ayurvedic literature and not in electronic form, hence making it is difficult to access. Many publications are not in peer-reviewed journals and the quality of available publications is questionable.
2. Most Ayurvedic formulations are multi-ingredient-fixed dose formulations rarely prescribed alone (i.e., there are multiple herbal and herbo-mineral FDCs being consumed at the same time).
3. Additionally, there is the confounding factor that the patient is often receiving allopathic medicines at the same time.
4. Pharmacokinetics and toxicokinetics are very difficult, and at this point of time, well nigh impossible making definite causality virtually impossible.
5. Dose-related responses are rarely measured and reported.
6. Rarely, if ever, is de-challenge and re-challenge performed and there is no objective evidence of the adverse event.
7. One of the most challenging aspects is the lack of expertise in performing causality analysis with Ayurvedic medicines. A person trained in pharmacovigilance rarely understands Ayurveda while an expert in Ayurveda is not trained in the science of pharmacovigilance.

7.4. PREVENTION OF ADVERSE REACTIONS TO AYURVEDIC MEDICINES

The success in any pharmacovigilance system is in the ability to prevent further adverse reactions successfully by understanding and using the information collected. With Ayurvedic medicines, the challenges would be at multiple levels.

1. Communication between the practitioners and policy makers of orthodox Western medicine and traditional Indian medicine is not adequate. In India, the current NPVP does not have Ayurveda under its fold and therefore Ayurvedic practitioners are not aware of the need to report and where to report.
2. Unbiased drug information about Ayurvedic drugs, including both classical and proprietary formulations is not available easily.
3. Patients are not adequately aware that Ayurvedic medicines can cause adverse reactions and can take medicines for years on end with no monitoring as they believe that these medicines can do no harm. Hence, they do not even give the history of taking these medicines to the practitioners.
4. Education in Ayurveda or modern medicine at both under-graduate and post-graduate levels do not cover pharmacovigilance of Ayurvedic medicines, thus never exposing the young physicians to this concept.
5. The Ayurvedic pharmaceutical industry is not motivated to focus on pharmacovigilance of Ayurvedic medicines. Hence, there is no attempt at generating safety data – either before or after marketing of the formulation.
6. Availability of Ayurvedic medicines is unprecedented in India! It is reported that there are over 100 books describing different Ayurvedic medicines containing over 100,000 recipes for medicines! The formal Ayurvedic formulary quotes over 630 formulations in its two published volumes. Add to that the huge informal sector, the numbers are mind boggling. The question arises then, which medicines should be included in the pharmacovigilance system?

7.5. PHARMACOVIGILANCE PROGRAM OF INDIA (PvPI)

The Pharmacovigilance Program of India (PvPI) was launched with a broad objective to safeguard the health of 1.27 billion people of India. Adverse Drug Reactions (ADRs) are reported from all over the country to NCC-PvPI, which also work in collaboration with the global ADR monitoring center (WHO-UMC), Sweden to contribute in the global ADRs data base. NCC-PvPI monitors the ADRs among Indian population and helps the regulatory authority of India (CDSCO) in taking decisions for the safe use of medicines.

7.6. NATIONAL PHARMACOVIGILANCE PROGRAM FOR AYURVEDA, SIDDHA AND UNANI DRUGS

The idea of a corresponding pharmacovigilance program for traditional medicine began in November 2006, principally led by the Department of Clinical Pharmacology, TNMC and BYL Nair Ch Hospital, Mumbai. In collaboration with WHO, clinical pharmacologist Urmilla Thatte and Vaidya Supriya Bhalerao organized a workshop, "Pharmacovigilance of Ayurvedic Medicine" on 20 and 21 November, 2006.

In October, 2007, BHU's Institute of Medical Sciences' Department of Rasa Shastra organized a seminar-cum-workshop entitled "Safety Profile of Ayurvedic Dosage Forms". The seminar's technical report submitted to WHO strongly influenced how the Ayurvedic medicines' pharmacovigilance safety program was implemented.

The next concrete step sponsored by WHO was taken by the Institute of Post Graduate Teaching and Research in Ayurveda (IPGTRA), at Gujarat Ayurveda University, Jamnagar, which organized a workshop on the possibility of implementing pharmacovigilance programs for Ayurvedic medicine in December, 2007. In view of the program's potential importance, the Dept. of AYUSH requested IPGTRA to prepare a protocol and ADR reporting format to implement pharmacovigilance for Ayurveda, Siddha, and Unani (ASU) drugs.

Pt. Khushilal Sharma Government Ayurveda College and Institute, Bhopal is working as the Regional Pharmacovigilance Centre of the central zone to collect and compile data of adverse drug reactions (ADRs) to Ayurveda, Siddha & Unani drugs, under the National Pharmacovigilance Resource Centre, IPGT&RA, Gujarat Ayurveda University, Jamnagar.

7.7. SET UP OF PHARMACOVIGILANCE FOR ASU DRUGS

Departments of government, hospitals, and academic institutions, involved in clinical pharmacology, clinical pharmacy, clinical toxicology, or epidemiology have been identified as the best hosts to set up and house centers. Those applying will be assessed for infrastructure and other resource requirements, since proper planning is essential to establish and run a pharmacovigilance center successfully. Also, government support is necessary at least at the national level.

At present, besides the National Center in Jamnagar, 8 Regional Centers and 30 Peripheral Centers are being developed to carry out program activities such as receiving reports submitted through the www.ayurveduniversity.com portal, where a complete list of centers may be found. Continuing Medical Education and public meetings are being conducted to raise health professionals' awareness of ASU drugs.

7.8. SCOPE OF PHARMACOVIGILANCE IN AYURVEDA

The goals of Ayurveda's pharmacovigilance program are to improve:

- Patient care and safety when using Ayurvedic medicines and related interventions;
- Public health and safety records of Ayurvedic medicines;
- Assessment of benefit, harm, effectiveness, and risk of medicines;
- Encouragement of safe, rational, and more effective (including cost effective) use, and promotion of understanding, education, and clinical training in pharmacovigilance for Ayurvedic medicines and its effective communication to the public. Many cases have been reported in the recent past regarding ADRs and drug-drug interactions at various national and international forums.

7.9. CENTRE FOR SAFETY & RATIONAL USE OF INDIAN SYSTEMS OF MEDICINE (CSRUISM)

WHO emphasizes that it should include traditional medicines in Pharmacovigilance system and has published guidelines on safety monitoring of herbal medicines in Pharmacovigilance systems in 2004. To promote the guidelines of WHO, Ibn Sina Academy of Medieval Medicine & Sciences, took a novel task of improving the use of Indian originated drugs and their adverse reaction monitoring under the establishment of Centre for Safety & Rational Use of Indian Systems of Medicine (CSRUISM) in 2005.

The Centre is also cooperating and assisting the National Pharmacovigilance Programme for Ayurveda, Siddha and Unani (ASU) Drugs, which is currently being governed under the aegis of Institute of Post Graduate Teaching & Research in Ayurveda, Gujarat Ayurveda University, Jamnagar in collaboration with the Department of AYUSH, Ministry of Health and FW, Govt. of India, New Delhi.

CHAPTER 8
HEAVY METAL CONTENT OF AYURVEDIC FORMULATIONS

8.1. HEAVY METALS

A heavy metal is one with a specific gravity of 5 gm cm^{-3} or more. They are stable elements and cannot be metabolized by the body. Some heavy metals like zinc, copper, cobalt, chromium, iron and manganese are required by the human body in small doses. Heavy elements can be toxic in larger quantities. Heavy elements may enter the body in food, water, air or by absorption through skin.

8.2. BHASMA OF METALS

Ayurvedic theory attributes important therapeutic roles of metals such as mercury and lead. Ayurveda experts estimate that 35% to 40% of the approximately 6000 medicines in the Ayurvedic formulary intentionally contain at least one metal. Metal-containing Ayurvedic herbal medicine products are purportedly "detoxified" through multiple heating/cooling cycles and the addition of specific herbs.

The heavy metals in elemental form are toxic, but those in compound form, i.e., *bhasma* is safe for humans. *Bhasma* are prepared from various metals, metal mixtures and alloys. According to Indian alchemy, *bhasma* prepared, evaluated and therapeutically used are absolutely safe to humans.

8.3. HEAVY METAL ANALYSES IN AYURVEDIC FORMULATIONS

Although adulteration with drugs is by definition fraudulent, the inclusion of heavy metals could be either intentional for alleged medicinal purposes or accidental. Evidence from various countries implies that toxic heavy metals and undeclared prescription drugs in Asian herbal medicines might constitute a serious health problem (Ernst 2002a).

One study showed that 64% of samples collected in India contained significant amounts of lead, mercury, arsenic and cadmium (Ernst 2002). A total of 14 (20%) of 70 HMPs (95% confidence interval, 11%–31%) contained lead, mercury, and/or arsenic (Saper et al. 2004).

Table 1. Heavy Metal Concentrations in Ayurvedic Medicinal Products (N = 70).*

	Metal							
	Lead	Mercury	Arsenic	Tin	Sliver	Gold	Cadmium	Any Metal*
No. of HMPs with detectable levels of metals	13(19)	6(9)	6(9)	5(7)	4(6)	3(4)	NA	14(20)
Median conc. among HMPs with detectable levels	40(8-300)	20225(4380-72100)	430(54-2800)	22800(3940-23500)	14310(2475-26575)		NA	

*A total number of 14 (20%) of 70 HMPs contained lead, mercury, and/or arsenic
Source: Saper et al. *JAMA* 2004; 292: 2868–2873.

One hundred ninety-three of the 230 requested medicines were received and analyzed. The prevalence of metal-containing products was 20.7% (95% confidence interval [CI], 15.2%–27.1%). The prevalence of metals in US-manufactured products was 21.7% (95% CI, 14.6%–30.4%) compared with 19.5% (95% CI, 11.3%–30.1%) in Indian products ($P = .86$). *Rasa shastra* compared with *non-rasa shastra* medicines had a greater prevalence of metals (40.6% vs. 17.1%; $P = .007$) and higher median concentrations of lead (11.5 µg/g vs. 7.0 µg/g; $P = .03$) and mercury (20800 µg/g vs. 34.5 µg/g; $P = .04$) (Saper et al. 2008).

A study used the Atomic Absorption Spectrometer for assessment of heavy metals, i.e., Zn^{2+}, Ni^{2+}, Pb^{2+} from Ayurvedic formulations. According to WHO and FDA, the average amount of zinc should be 0.6 mg/kg, lead should be 100 mg/kg and nickel should be 50 mg/kg (Singh 2008). These values may vary due to environmental factors. The present study showed that the amount of zinc was more in *Mahayograj Guggul* and *Mahalaxmivilas Ras* than the WHO and FDA report. The amount of lead and nickel was found more than the WHO and FDA report in *Mahayograj Guggul*. These excess amounts of heavy metals in Ayurvedic drugs are due to a manufacturing defect, which is responsible for causing genotoxicity (Table 1).

Table 2. Amount of heavy metals in the sampled Ayurvedic drugs.

Formulation	Zn (mg/kg)	Pb (mg/kg)	Ni (mg/kg)
Mahayograj Guggul	176.2	112.6	75.2
Mahalaxmivilas Ras	171.1	98.6	32.1

8.4. DEPARTMENT OF AYUSH AND HEAVY METALS IN AYURVEDIC FORMULATIONS

Saper et al. 2004 published an article in JAMA on "Heavy Metal Content of Ayurvedic Herbal Medicine Products" on the basis of testing of 70 samples of herbal medicinal products collected from grocery stores in Boston Area for heavy metals. Department of AYUSH enforced mandatory testing for heavy metals in respect of Ayurveda, Siddha and Unani herbal products being exported from India w.e.f. 1 January 2006. Testing for heavy metals and other contaminants in Ayurveda, Siddha and Unani products is already a part of Good Manufacturing Practices notified in 2003.

A project for physicochemical characterization and toxicity studies of eight widely used Bhasmas (Rasa Aushadhies) was also sanctioned under the Golden Triangle Project, which is being carried out by the various laboratories of CSIR, i.e., Indian Institute of Toxicological Research (IITR), Lucknow, Indian Institute of Chemical Technology (IICT), Hyderabad. Under this project, one of the reputed manufacturers of Rasa Aushadhies was chosen for manufacturing of the selected Rasa Aushadhies as per the classical texts which were made available to CSIR laboratories for physiochemical characterization and their toxicity studies. On the basis of 28 day toxicity studies, all the eight Rasa Aushadhies have been found to be non-toxic. 90 days chronic studies are under progress. Further, the work of finalizing SOPs for the various herbo metallic compounds (Rasa Aushadhies) used in Ayurveda has been undertaken by the Ayurvedic Pharmacopoeia Committee of India. Supplementary Good Manufacturing Practices for Rasa Aushadhies have also been prepared of which draft publication has been done.

Samples of 600 Indian medicinal plants collected from the wild as well as various medicinal plant gardens in India by the Council for Scientific Research in Ayurveda and Siddha were sent to the Indian Institute of Toxicological Research (CSIR), Lucknow, Sri Ram Institute of Industrial Toxicology, New Delhi and Centre for Research in Indian Medicine, Shastra University, Thanjavur. The test reports received from these three laboratories disclose that lead, mercury and arsenic have not been found in these 600 Indian medicinal plant samples above the permissible limits laid down by WHO which is 10 ppm for Lead, 1 ppm for Mercury and 3 ppm for Arsenic.

It needs to be emphasized that as per the directions issued by Department of AYUSH, herbo metallic compounds are not being officially exported because of heavy metal concerns and only purely herbal Ayurveda, Unani and Siddha medicines are being exported from India with effect from 1 January 2006 after certification of heavy metals below the permissible limit by the manufacturing unit. In view of the above, the above mentioned article of Dr. Saper and his associates is seriously flawed and discloses a strong bias against Ayurvedic medicines. Indian scientists and research institutions will be responding to the issues raised by Dr. Saper, howsoever flawed they may be, through research articles based on their work on Ayurvedic medicines in due course.

8.5. LEAD POISONING FROM AYURVEDIC MEDICINES

A case regarding lead poisoning caused by Indian ethnic remedies in Italy was reported (Muzi et al. 2005). A 32-year-old man was repeatedly hospitalized for paroxysmal abdominal pain with constipation, weight loss, anemia, and mild elevation of liver enzyme levels. Four months after the initial admission, blood lead measurement showed heavy metal poisoning (Garnier and Poupon 2006).

A 60-year-old man with a history of diabetes and hypertension was referred to a nephrology clinic for investigation of his elevated serum creatinine level. The patient was diagnosed with Stage 3 chronic kidney disease, probably which was worsened by consumption of lead in the form of an Ayurvedic herbal remedy (Prakash et al. 2009). A 58-year-old type II diabetic man took Ayurvedic medication for generalized weakness and developed peripheral neuropathy following its intake. He was found to have high blood and urinary lead levels and was diagnosed to have sub acute lead poisoning (Singh et al. 2009). A 58-year-old woman consuming Jambrulin complained with abdominal pain, anemia, liver function abnormalities and an elevated blood lead level. Chemical analysis of the medication showed high levels of lead (Gunturu et al. 2011).

The presence of free lead in five different formulations indicated towards the possible risk of severe side effects to the consumer. Present findings certainly put doubt over the safety of this formulation, but at the same time, variation in the results with all five formulations also indicated that these formulations were not prepared as per the mentioned Ayurvedic text (Garg et al. 2012). The elemental characterization of preparations containing *Naga bhasma* has shown extremely high levels of lead content and various parameters must be taken into consideration in deciding the safety and critical issues present in traditional medicines (Nagarajan et al. 2014).

8.6. ARSENIC POISONING FROM AYURVEDIC MEDICINES

Consumption of the hepatotoxin arsenic is very common in certain geographical areas of India and occurs as a result of the intake of arsenic contaminated water, vegetables, adultered opium, Ayurvedic and indigenous medicines and "home-made brew" (Narang 1987).

An 11-year-old girl developed manifestations of arsenical keratosis (punctuate palmoplantar keratoderma and leucomelanoderma) and non-cirrhotic portal hypertension, 6 months and 18 months respectively after intake of Ayurvedic medications, prescribed for epilepsy. Arsenic content of Ayurvedic medicines ranged from 5 mg/L to 248 mg/L. The serum arsenic level was 202.20 microg/L (normal < 60 microg/L) (Khandpur et al. 2008).

Manufacture of an Ayurvedic arsenic-containing compound is described, which is currently in use in India to control blood counts of patients with haematological malignancies. The efficacy and side effects of this compound are evaluated in the light of the fact that arsenic was recognised to be of use in the control of blood counts from patients with chronic myeloid leukaemia as long as 100 years ago in the West (Treleaven et al. 1993). Toxicology studies on arsenic formulations used in Ayurveda (*Haritala, Manashila* and *Gauripashana* are tabulated below).

Table 3. Toxicology studies on arsenic formulations used in Ayurveda.

Formulation	Short-term Toxicity	Long-term Toxicity	Remark
Haritala	LD_{50} = 6.4 gm/kg	Negative in micro-nucleus assay	High level in blood
Manashila	LD_{50} = 3.2 gm/kg	Skin manifestations, mild GI discomfort, fatty lover and prolonged QT	Adverse effect tolerable
Gauripashana	LD_{50} = 32.29 gm/kg	Heart, skin and GI effects	Dose-dependent, tolerable

Source: Panda and Hazra. *IJRAP* 2012; 3: 772–6.

8.7. ASSESSMENT OF HEAVY METAL CONTENTS OF SOME INDIAN MEDICINAL PLANTS

The present study was carried out to quantitatively analyze the levels of six potentially toxic heavy metals viz., Arsenic (As), lead (Pb), cadmium (Cd), mercury (Hg), chromium (Cr) and nickel (Ni) in ten important Indian medicinal plants. The air dried, powdered plant materials were subjected to microwave assisted wet digestion for the preparation of test samples. The samples were analyzed by using atomic absorption spectrophotometer equipped with graphite tube atomizer. From the results of the present investigation, it can be inferred that the levels of all six hazardous heavy metals were below the permissible limits in all ten medicinal plants analyzed (Singh, Bhattacharya and Sharma 2014).

Table 4. Heavy metal contents in studied medicinal plants (ppm).

S. No.	Samples	As	Pb	Cd	Hg	Cr	Ni
1.	*C. pluricaulis*	0.38	1.03	BLQ	BLQ	1.11	0.44
2.	*T. belerica*	0.46	1.22	0.06	BLQ	1.88	0.53
3.	*T. chebula*	0.09	0.93	BLQ	BLQ	1.19	0.64
4.	*S. aromaticum*	0.18	0.98	0.08	BLQ	1.74	0.87
5.	*S. nux-vomica*	BLQ	1.16	BLQ	BLQ	1.91	0.8
6.	*A. vasica*	0.35	0.72	0.07	0.08	0.79	0.88
7.	*O. sanctum*	0.14	0.89	BLQ	BLQ	1.58	0.65
8.	*C. fistula*	BLQ	1.28	BLQ	BLQ	1.28	0.47
9.	*E. alba*	0.08	0.86	0.05	BLQ	1.56	0.42
10.	*C. zeylanicum*	0.43	0.69	BLQ	BLQ	1.63	0.43

BLQ: Below limit of quantification, i.e., 0.05 ppm
Source: Singh, Bhattacharya and Sharma. *American-Eurasian J Agric Environ Sci* 2014; 14: 1125–9.

8.8. TOXICOLOGICAL STANDARDIZATION OF MARKETED ASHWAGANDHA FORMULATIONS BY ATOMIC ABSORPTION SPECTROSCOPY

Atomic absorption spectrometry is an advanced technique to ascertain accumulation of heavy metals including arsenic, cadmium and lead in Herbal formulations. In this study, the marketed Ashwagandha formulations Yavatmal City (India), were investigated by this technique. The main purpose of the investigation was to document evidence for the users, and practitioners of marketed Ashwagandha formulations. WHO (1998) mentions the maximum permissible limits in raw materials only for arsenic, cadmium and lead, which amount to 1.0, 0.3 and 10 ppm, respectively.

It was found that the arsenic content in herbal formulations was below the permissible limit in all formulations. The cadmium content in H2 (1.2 ppm), H3 (0.9 ppm), H4 (0.7 ppm), H5 (0.93 ppm), H7 (1.1 ppm), H8 (0.56 ppm), H9 (0.75 ppm) and H10 (0.34 ppm) were above the permissible limits. The lead content in H2 (15.5 ppm), H5 (12.5 ppm), H6 (11.7 ppm), H7 (12.9 ppm) and H9 (15.9 ppm) were above the permissible limits. Such formulations may cause damage to the delicate organs of patient as they get accumulated in the body (Bais and Chandewa 2013).

Table 5. Arsenic content in Ashwagandha formulations code H1 to H10.

Formulation Code	Arsenic Content (ppm)	Remark
H1	0.02	Within permissible limit
H2	0.04	Within permissible limit
H3	0.05	Within permissible limit
H4	0.021	Within permissible limit
H5	0.06	Within permissible limit
H6	0.028	Within permissible limit
H7	0.056	Within permissible limit
H8	0.034	Within permissible limit
H9	0.06	Within permissible limit
H10	0.073	Within permissible limit

Source: Bais and Chandewa *Asian J Pharm Clin Res* 2013; 6: 45–8.

Table 6. Lead content in Ashwagandha formulations code H1 to H10.

Formulation Code	Arsenic Content (ppm)	Remark
H1	6.2	Within permissible limit
H2	15.5	Above permissible limit
H3	6.3	Above permissible limit
H4	4.8	Above permissible limit
H5	12.5	Above permissible limit
H6	11.7	Within permissible limit
H7	12.9	Above permissible limit
H8	1.9	Above permissible limit
H9	15.9	Above permissible limit
H10	3.2	Above permissible limit

Source: Bais and Chandewa *Asian J Pharm Clin Res* 2013; 6: 45–8.

8.9. ESTIMATION OF HEAVY METALS IN *ECLIPTA ALBA*

The present estimation showed that *E. alba* is suitable for the control of environmental pollutants, such as heavy metals. However for medicinal purposes, it should be collected from those areas, which are not contaminated with heavy metals (Kinthada, Naidu and Muralidhar 2011).

Table 7. Concentration of heavy metal in plant materials and soil samples.

Selected Areas	Plant Material and Soil	Heavy Metals Concentration (mg/kg)*					
		Fe	Zn	Mn	Cr	Pb	Cu
Sample-1 Most polluted area Gajuwaka industrial zone in Visakhapatnam, A.P. India	Leaves Soil	23.9 27.9	1.61 4.02	6.97 6.47	0.47 0.55	0.40 0.52	1.18 0.33
Sample-2 Unpolluted area K. Palem (Village) Nellore-Dist. A.P. India	Leaves Soil	15.2 18.8	1.29 1.02	5.24 0.44	0.27 0.34	0.19 0.16	1.28 0.36

Source: Kinthada, Naidu and Muralidhar, *Asian J Pharm Clin Res* 2011; 2: 99-102-8.

8.10. ESTIMATION OF HEAVY METALS IN *MAHAYOGRAJ GUGGULU*

Heavy metal content determined using an atomic absorption spectrophotometer are presented in Table 8. The corresponding values in our study are 25.800 µg/g for lead, 0.07 µg/g for mercury and 5.19 µg/g for arsenic. The comparison shows that the values reported by Saper et al. are higher in comparison with the values reported by Central Council for Research in Ayurvedic Sciences (CCRAS). The values for lead alone are higher than those prescribed by WHO. The other values are within the prescribed limits.

Table 8. Heavy metal estimation in *Mahayograj guggulu*.

Name of the Organization	Pb	Hg	As	Cd
CCRAS*	25.800 µg/g	0.007 µg/g	5.19 µg/g	0.947 µg/g
JAMA**	37.000 µg/g	22.80 µg/g	8.100 µg/g	-

*Estimation by atomic absorption spectrophotometer
**Estimation by X-ray fluorescence spectroscopy
Source: Lavekar et al. *Int J Ayurveda Res* 2010; 1: 150–8.

8.11. ESTIMATION OF HEAVY METALS IN *NAVRATNA RASA*

To determine the metal concentration in *Navratna rasa*, the sample was analyzed by the Atomic Absorption Spectrophotometer. The sample was weighed and taken in a porcelain basin and ashed in a muffle furnace at 525°C. The ash was extracted with 1:1 HCl solution and then volume were measured. Heavy metals of the extracted solution were estimated with an Atomic Absorption Spectrophotometer. For estimation of arsenic, Hydride Generator was used (Lavekar et al. 2009).

Table 9. Heavy metal estimation in *Navratna rasa*.

Name of the Organization	Pb	Hg	As	Cd
CCRAS*	600 µg/g	10400 µg/g	60 µg/g	-
JAMA**	47.05 ppm	0.05 ppm	4.18 ppm	0.48 ppm

*Estimation by X-ray fluorescence spectroscopy
**Estimation by atomic absorption spectrophotometer
Source: Lavekar et al. *Toxicol Int* 2009; 16: 37–42.

8.12. ESTIMATION OF HEAVY METALS IN *CHANDRAPRABHA VATI*

Table 10. Heavy metal estimation in *Chandraprabha vati*.

Sample Code	Hg (%)	Pb (ppm)	Cd (ppm)
Marketed sample	0.33	69.31	BDL*
In-house preparation	0.41	BDL	BDL

*BDL – Below Detectable Limits
Source: Saralla et al. *Int J Pharm Pharm Sci* 2013; 6: 80–4.

CHAPTER 9

GENOTOXIC POTENTIAL OF AYURVEDIC FORMULATIONS

9.1. GENOTOXICITY

Genotoxicity describes the property of chemical agents that damages the genetic information within a cell causing mutation, which may lead to cancer. The alteration can have direct or indirect effects on the DNA: the induction of mutations, mistimed event activation and direct DNA damage leading to mutations.

9.2. GENOTOXICITY OF AYURVEDIC FORMULATIONS

Ras Manikya Ras, *Lauha Bhasma*, *Tamra Bhasma* and *Kajjali Bhasma* were investigated for genotoxic potential. A single dose (200 mg/kg b.w.) was administered orally to Wistar rats. Peripheral blood leukocytes and bone marrow samples were collected. Micronucleus assay and the comet assay were employed to study the endpoint of chromosomal damage and single/double-strand DNA breaks. The results revealed a lack of induction of micronuclei or DNA damage as evidenced by the Comet assay, despite the presence of traces of transformed toxic heavy metals (Sathya et al. 2008).

Abhraka Bhasma, *Mandura Bhasma* and herbomineral formulations such as *Swasa Kuthara rasa* and *Smriti Sagara rasa* were reported to be free from genotoxicity (Vardhini et al. 2010).

The micro-nucleus frequency in *Mahayograj Guggul* treated lymphocytes varies from 3.8 to 4.0% and 3.4 to 3.6% in *Mahalaxmivilas Ras* treated lymphocytes. The result indicates that *Mahayograj Guggul* is the most genotoxic because of the presence of high amount of heavy metal and *Mahalaxmivilas Ras* is least genotoxic because of a low amount of heavy metal (Mishra 2014).

CHAPTER 10

ARISTOLOCHIC ACID DISTRIBUTION IN AYURVEDIC FORMULATIONS

10.1. ARISTOLOCHIC ACID

Aristolochic acids are a family of carcinogenic, mutagenic, and nephrotoxic compounds commonly found in the plant family Aristolochiaceae, including *Aristolochia* and *Asarum* (wild ginger), which are commonly used Chinese herbal medicine (Heinrich et al. 2009). Aristolochic acid is composed of a ~1:1 mixture of two forms, aristolochic acid I and aristolochic acid II. Aristolochic acid I is the most abundant of the aristolochic acids and is found in almost all Aristolochia species (Wu et al. 2005). Aristolochic acids are often accompanied by aristolactams.

10.2. TOXICITY

Aristolochic acids are hypothesized to be causative agents in Balkan endemic nephropathy (Gluhovschi et al. 2011) and a related—possibly identical—condition known as "Chinese herbs nephropathy" (De Broe 2012). Exposure to aristolochic acid is linked to urothelial cancer (Lai et al. 2009). However, despite these well-documented dangers, aristolochic acid is still present sometimes in herbal remedies (such as for weight loss), primarily because of the substitution of innocuous herbs with *Aristolochia* species (National toxicological program 2008).

10.3. FDA ALERT ON BOTANICAL PRODUCTS INCLUDING ARISTOLOCHIC ACID

FDA has issued a Consumer Advisory and sent updated letters to industry and health professionals to communicate our concern about the use and marketing of dietary supplements or other botanical-containing products that may contain aristolochic acid. The use of products containing aristolochic acid, including botanical products marketed as traditional medicines, has been associated with nephropathy.

10.4. AYURVEDIC PLANTS CONTAINING ARISTOLOCHIC ACID

Chronic interstitial nephropathies in Indians were mentioned to be associated with Aristolochia sp. (Vanherweghem 1997).

Aristolochia bracteata Retz. (Bracteated birthwort)

In Ayurveda, *A. bracteata* is known as *Kitmari* or *Dhumapatra*. It is a shrub distributed throughout India. In the indigenous system of medicine, the plant was used for the treatment of skin diseases, inflammation and as a purgative.

Aristolochia indica L. (Indian birthwort)

In Ayurveda, *A. indica* is known as *Jata*. *A. indica* is a creeper plant found in Kerala in India and also Sri Lanka. It is used in traditional medicine for postpartum infections and snakebite respectively (Rajashekharan et al. 1989).

Chronic interstitial nephropathies in Indians were mentioned to be associated with Aristolochia sp. (Vanherweghem 1997). This species was treated among known or suspected botanicals to contain aristolochic acid (Debelle et al. 2008).

10.5. ARISTOLOCHIC ACID NEPHROPATHY IN DEVELOPING COUNTRIES

Two cases of rapidly progressive fibrosing interstitial nephritis were reported in young women. It was later found that one herb in a weight-reducing formula (Stephania tetrandra) had been inadvertently replaced by *Aristolochia fangchi* by the suppliers (Vanherweghem et al. 1993).

Several commercial dietary supplements, teas and phytomedicines used as slimming regimens were analysed for their aristolochic acid I content. Out of 42 analysed preparations, four were found to contain aristolochic acid I and two were suspected to contain aristolochic acid derivatives. Immediate removal of these products from the Swiss market was called for (Ioset et al. 2003).

Species of Aristolochia are used medicinally in many regions of the world and both from an ethnopharmacological and a public health perspective this poses a risk. In China and Europe, species of Aristolochia have been associated with nephropathy and it is important to evaluate whether nephropathy occurs in other parts of the world, especially India and Central America where the use of species of Aristolochia are reported to be commonly used in traditional medicine (Heinrich et al. 2009).

Aristolochic acid nephropathy has been reported in ten countries, but its true incidence is unknown and most likely underestimated. By combining an ethnobotanical and phytochemical approach, researchers provided evidence for the risk of aristolochic acid nephropathy occurring in Bangladesh.

Knowledge about toxicity or side effects of *A. indica* is very limited and *A. indica* is often administered in very high doses. Replacement of *A. indica* with *Rauvolfia serpentina* (L.) Benth. ex Kurz was common. *A. indica* samples contained a variety of aristolochic acid analogues such as aristolochic acid I, aristolochic acid II, cepharadione A and related compounds (Michl et al. 2013).

CHAPTER 11
CLINICAL TRIALS IN AYURVEDA

The department of Ayush has issued good clinical practice (GCP) guidelines for clinical trials in Ayurveda, Siddha and Unani (ASU) medicines which will facilitate the researchers and institutions in adopting a standard way of good clinical practice while conducting the ASU clinical trials.

The guidelines are addressed to investigators and all those who are interested, concerned, involved and affected by the conduct of clinical trials on ASU drugs. These are timely in view of the focus being given for scientific validation and for promoting evidence-based use of ASU treatments and are meant for voluntary use, not linked with any provisions of Drugs & Cosmetics (D&C) Act, 1940, and the rules thereunder.

The objective of this chapter is to encourage the belief that clinical studies in ASU systems are undertaken in accordance with ethical and scientific standards and safety aspects and rights of participants are protected. Adhering to methodical documentation of trials will help bring credibility to the efforts of persons and institutions involved in the process, which otherwise was lacking for want of any ASU-specific guiding document.

11.1. GOOD CLINICAL PRACTICE GUIDELINES

The history of Good Clinical Practice (GCP) statute traces back to one of the oldest enduring traditions in the history of medicine. Ayurveda deeply emphasizes ethical guidelines while treating a patient through medical/surgical interventions. Utmost priority has been accorded to ethical issues and prior consent of the patient was suggested in the Ayurvedic texts.

The aim of Good Clinical Practice is to ensure that the studies are scientifically and ethically sound, and that the clinical properties of the ASU medicine under investigation are properly documented. The guidelines seek to establish two cardinal principles: protection of the rights of human subjects and authenticity of ASU medicine clinical trial data generated.

With the introduction of Drugs & Cosmetics Rule 158 B since August 2010, the requirement of proof of effectiveness for licensing of patent or proprietary ASU medicine has necessitated the development of present guidelines of Good Clinical Practice. However, these guidelines are for voluntary use by the researchers interested in taking up clinical trials by using ASU medicine. Conducting clinical trials and generating evidence on the basis of these guidelines would help convince the world about the potential scope of ASU remedies in scientific parlance and address the questions of lack of evidence and validation. Immense opportunities thus lie ahead for the stakeholders to adopt the guidelines as a tool for promoting scientific and quality clinical research for credible outcomes.

11.2. DEFINITIONS

11.2.1. Act

Wherever relevant, the Act means Drugs & Cosmetics Act, 1940 (23 of 1940) and the Rules made there under.

11.2.2. Adverse Event (AE)

Any untoward medical occurrence (including a symptom/disease or an abnormal laboratory finding) during treatment with a pharmaceutical product in a patient or a human volunteer that does not necessarily have a relationship with the treatment being given. Also see Serious Adverse Event.

11.2.3. Adverse Drug Reaction (ADR)

a) In case of approved ASU Medicines: A noxious and unintended response at doses normally used or tested in humans.
b) In case of new unregistered ASU Medicines: A noxious and unintended response at any dose(s).

The phrase ADR differs from AE. In case of an ADR, there appears to be a reasonable possibility that the adverse event is related to the medicinal product being studied.

In clinical trials, an untoward medical occurrence seemingly caused by overdosing, abuse/dependence and interactions with other medicinal products is also considered as an ADR.

Adverse drug reactions are type A (pharmacological) or type B (idiosyncratic). Type A reactions represent an augmentation of the pharmacological actions of a drug. They are dose dependent and are, therefore, readily reversible on reducing the dose or withdrawing the drug. In contrast, type B adverse reactions are bizarre and cannot be predicted from the known pharmacology of the drug.

11.2.4. Audit of a Trial

A systematic verification of the study, carried out by persons not directly involved, such as:
a) Study related activities to determine consistency with the Protocol.
b) Study data to ensure that there are no contradictions in Source Documents. The audit should also compare data on the Source Documents with the interim or final report. It should also aim to find out if practices were employed in the development of data that would impair their validity.
c) Compliance with the adopted Standard Operating Procedures (SOPs).

11.2.5. Ayurveda, Siddha and Unani Drugs

"Ayurvedic, Siddha or Unani drug" includes all medicines intended for internal or external use for or in the diagnosis, treatment, mitigation or prevention of disease or disorder in human beings or animals, and manufactured exclusively in accordance with the formulae described in, the authoritative books of Ayurvedic, Siddha and Unani Tibb system of medicine, specified in the First Schedule.

11.2.6. Patent or Proprietary Medicine

In relation to Ayurvedic, Siddha or Unani Tibb systems of medicine, all formulations containing only such ingredients mentioned in the formulae described in the authoritative books of Ayurveda, Siddha or Unani Tibb system of medicine specified in the First Schedule, but does not include a medicines which is administered by parenteral route and also a formulation included in the authoritative books as specified in clause (a).

11.2.7. Blinding/Masking

A method of "control experimentation" in which one or more parties involved are not informed of the treatment being given. Single blind refers to the study subject(s) being unaware, while Double blind refers to the study subject(s) and/or investigator(s), monitor, data analyst(s) being unaware of the treatment assigned.

11.2.8. Case Record Form (CRF)

A document designed in consonance with the Protocol, to record data and other information on each trial subject. The Case Record Form should be in such a form and format that allows accurate input, presentation, verification, audit and inspection of the recorded data. A CRF may be in printed or electronic format.

11.2.9. Clinical Trial (Clinical Study)

A systematic study of ASU Medicines on human subjects – (whether patients or non-patient volunteers) – in order to discover or verify the clinical, pharmacological (including pharmacodynamics/pharmacokinetics), and/or adverse effects, with the object of determining their safety and/or efficacy.

11.3. PHASES OF CLINICAL TRIAL FOR ASU DRUGS

11.3.1. Human Pharmacology (Phase I)

i) The objective of studies in this Phase is the estimation of safety and tolerability with the initial administration of ASU Drugs/other T M new drugs into human(s). Studies in this phase of development usually have non-therapeutic objectives and may be conducted in healthy volunteer subjects or certain types of patients. Drugs with probable toxicity, e.g., drugs with Schedule E-I ingredients are usually studied in patients. Phase I trials should preferably be carried out with access to the necessary facilities to closely observe and monitor the subjects.

ii) Studies conducted in Phase I, usually intended to involve one or a combination of the following objectives:

iii) Maximum tolerated dose: To determine the tolerability of the dose range expected to be needed for later clinical studies and to determine the nature of adverse reactions that can be expected. These studies include both single and multiple dose administration.

iv) Early measurement of drug activity: Preliminary studies of activity or potential therapeutic benefit may be conducted in Phase I as a secondary objective. Such studies are generally performed in later Phases but may be appropriate when drug activity is readily measurable with a short duration of drug exposure in patients at this early stage.

11.3.2. Therapeutic Exploratory Trials (Phase II)

i) The primary objective of Phase II trials is to evaluate the effectiveness of an ASU drug for a particular indication or indications in patients with the condition under study and to determine the common short-term side-effects and risks associated with the drug. Studies in Phase II should be conducted in a group of patients who are selected by relatively narrow criteria leading to a relatively homogeneous population. These studies should be closely monitored. An important goal for this Phase is to determine the dose(s) and regimen for Phase III trials. Doses used in Phase II are usually (but not always) less than the highest doses used in Phase I. These studies should be intended to provide an adequate basis for marketing approval for ASU Drugs.

ii) Additional objectives of Phase II studies can include evaluation of potential study endpoints, therapeutic regimens (including concomitant medications) and target populations (e.g., mild versus severe disease) for further studies in Phase II or III.

iii) These objectives may be served by exploratory analyses, examining subsets of data and by including multiple endpoints in trials.

iv) If the application is for conduct of clinical trials as a part of multi-national clinical development of the drug, the number of sites and the patients as well as the justification for undertaking such trials in India shall be provided to the Licensing Authority.

11.3.3. Therapeutic Confirmatory Trials (Phase III)

i) Phase III studies have the primary objective of demonstration or confirmation of therapeutic benefits(s). Studies in Phase III are designed to confirm the preliminary evidence accumulated in Phase II that a drug is safe and effective for use in the intended indication and recipient population. Studies in Phase III may also further explore the dose-response relationships (relationships among dose, and clinical response), use of the drug in wider 12 populations in different stages of disease or the safety and efficacy of the drug in combination with other drug(s).
ii) For drugs intended to be administered for long periods, trials involving extended exposure to the drug are ordinarily conducted in Phase III, although they may be initiated in Phase II. These studies carried out in Phase III complete the information needed to support adequate instructions for use of the drug (prescribing information).
iii) For ASU drugs approved outside India, Phase III studies need to be carried out primarily to generate evidence of efficacy and safety of the drug in Indian patients when used as recommended in the prescribing information. Prior to conduct of Phase III studies in Indian subjects, Licensing Authority may require detailed safety studies and if possible, pharmacokinetic studies to be undertaken to verify that the data generated in the Indian population is in conformity with the data already generated abroad.
iv) If the application is for the conduct of clinical trials as a part of multi-national clinical development of the drug, the number of sites and patients as well as the justification for undertaking such trials in India should be provided to the Licensing Authority along with the application.

11.3.4. Post Marketing Trials (Phase IV)

Post Marketing trials are studies (other than routine surveillance) performed after drug approval and related to the approved indication(s). These trials go beyond the prior demonstration of the drug's safety, efficacy and dose definition. These trials may not be considered necessary at the time of new drug approval, but may be required by the Licensing Authority for optimizing the drug's use. They may be of any type, but should have valid scientific objectives. Phase IV trials include additional drug-drug interaction(s), dose response or safety studies and trials designed to support use under the approved indication(s), e.g., mortality/morbidity studies, epidemiological studies, etc.

11.4. STANDARDS OF REPORTING AYURVEDIC CLINICAL TRIALS

Reported lack of efficacy of Ayurvedic treatments in clinical trials is often not due to inefficacy of the treatment itself, but arises from inadequacies of trial design. The discussion argues that the trials of Ayurvedic interventions should exclusively use its multi-component, individualized and inherently holistic approach, and that the general guidelines for rigorous reporting of such clinical trials should be developed. Holistic Ayurvedic clinical trials, rigorously conducted and with high standards of reporting should translate into *good clinical science*, and may be expected to generate higher credibility for clinical studies of the Ayurvedic knowledge system (Mathur and Sankar 2010).

CHAPTER 12
NATIONAL AYUSH MISSION

12.1. DEPARTMENT OF AYUSH

Department of Indian Systems of Medicine and Homoeopathy (ISM&H) was created in March, 1995 and re-named as Department of Ayurveda, Yoga & Naturopathy, Unani, Siddha and Homoeopathy (Ayush) in November, 2003 with a view to providing focused attention on Development of Education and Research in these systems of medicine. The Department continues to lay emphasis on upgradation of Ayush educational standards, quality control and standardization of drugs, improving the availability of medicinal plant material, research and development and awareness generation about the efficacy and safety of the systems domestically and internationally.

12.2. NATIONAL POLICY ON AYUSH

National Health Policy 1983 refers to our rich, centuries old heritage of medical and health sciences. The Policy reveals that although vast infrastructure is available in the Indian Systems of Medicine and Homoeopathy for addressing health care of our people, they are underutilized.

The Central Council for Health and Family Welfare in 1999 recommended, inter-alia, that at least one physician from the Indian Systems of Medicine and Homoeopathy should be available in every Primary Health Care Centre (PHC) and that vacancies caused by non-availability of allopathic personnel should be filled by ISM&H physicians. The Council also resolved that specialist ISM&H treatment centers should be introduced in rural hospitals and a wing should be created in existing state and district level government hospitals to extend the benefits of these systems to the public.

12.3. WHY NATIONAL AYUSH MISSION?

India possesses an unmatched heritage represented by its ancient systems of medicine like Ayurveda, Siddha, Unani and Homoeopathy which are a treasure house of knowledge of preventive and promotive healthcare. The positive features of the Indian systems of medicine, namely their diversity and flexibility; accessibility; affordability, a broad acceptance by a large section of the general public; comparatively lesser cost and growing economic value, have great potential to make them providers of healthcare that the large sections of our people need.

12.4. OBJECTIVES OF NATIONAL AYUSH MISSION

A. The improvement of Ayush education through enhancement in the number of upgrading educational institutions;
B. Better access to Ayush services through increase in the number of Ayush hospitals and dispensaries, availability of drugs and manpower;
C. Providing sustained availability of quality raw material for Ayush systems of medicine; and
D. Improving availability of quality ASU&H drugs through increase in the number of pharmacies, drug laboratories and improved enforcement mechanism of ASU&H drugs.

12.5. ISSUE OF IMPROVEMENT OF AYUSH EDUCATION

Courses in Ayush are available at undergraduate (UG) and postgraduate (PG) levels. Central Council of Indian Medicine (CCIM) is the statutory, regulatory body constituted under IMCC Act 1970 regulating the Ayurveda education at degree level as well as at postgraduate level as well as the practice of Ayurvedic medicine in India. Central Council of Homeopathy is a statutory body of the Government of India under the Ministry of Health & Family Welfare, Department of Ayush for regulating homoeopathic education.

In AYUSH, recognition of medical qualifications is essential for shaping the carrier career of professionals. There is a second Schedule to the Indian Medicine Central Council Act, 1970 in which recognized medical qualification of Ayurveda/Siddha/Unani is included. The Government of India amends this Schedule from time to time by including medical qualifications of Indian Systems of Medicine by issuing a Gazette notification. Degrees awarded by universities fall in following categories:

1. Recognised: If the course is opened with Government of India permission.
2. Illegal: If the course is opened without Government of India permission, then it is an illegal one and such act is also punishable. No one can use this degree for employment or even display it as a qualification.
3. Unrecognised: If such course is opened after the Government of India permission, but there are deficiencies as per CCIM requirement or CCIM inspection process is underway, then it is unrecognized but may get recognized once all things are in place.
4. Derecognised: The courses were opened with permission of the Government of India and earlier were recognized by CCIM also but later on CCIM recommended/ declared them as derecognised course on their subsequent inspecting. Such courses may also get recognised again.

The factual position of system of Schedule in AYUSH qualifications is in fact confusing. A student before taking admission in Ayush institution is unaware of the recognition/unrecognition status of the UG/PG course. The student is neither eligible for Central/State registration nor for application in government sector.

If improvement of AYUSH education mandate of the National AYUSH Mission has to be achieved, we need to address human resource issues in AYUSH sector. Either the system of Schedule in AYUSH qualifications should be abolished or steps should be initiated for including all the Ayush qualifications in recognised category.

Recently B.A.M.S. qualification awarded by Guru Ravi Dass Ayurvedic University, Hoshiarpur has been brought under the umbrella of recognised qualifications by amendment of the 2nd Schedule. Similarly B.A.M.S. and M.D. (Ayurveda) qualifications awarded by Vikram University, Ujjain was included in the 2nd Schedule of their IMCC Act, 1970 thereby protecting future of pass outs.

On the other hand, several qualifications were not included in the 2nd Schedule of the IMCC Act, 1970 thereby putting a question mark over the future of pass outs. As an instance M.D. (Ayurveda) qualifications awarded by BRA Bihar University, Muzaffarpur stands in an unrecognised category despite its initiation in 1977–78.

12.6. FUTURE PROSPECTS

Bhore Committee (1945) recommended the establishment of a chair of history of Medicine in the All India Medical Institute. Government of India decided to implement some of the recommendations of Chopra Committee (1946). As a result of the Pandit Committee's report, the Central Research Institute in indigenous systems of medicine was created in Jamnagar in 1952. Like Chopra Committee, Government of India decided to implement some recommendations of the Dave Committee (1955). Udupa Committee (1958) led reforms of Indian Systems of Medicine (ISM), including changes in education and research at the national level.

National Policy on AYUSH followed by National AYUSH Mission are welcome steps for promoting Ayush. However, the bitter truth is that despite several announcements by the Government, things have failed to take practical shape for AYUSH. It is sincerely hoped that C.C.I.M. will play a pivotal role in implementation of efforts initiated by the creation of National AYUSH Mission.

CHAPTER 13
AYURVEDIC PHARMACY EDUCATION

13.1. INTRODUCTION

Ayurvedic pharmacy courses taught in traditional mode is known as Upvaid. The term highlights Ayurvedic dispenser or pharmacist. The course so far has remained in hibernation either due to non availability of job prospects or non-popularity of Ayurveda. Recently, Ayurvedic drug industry has witnessed an increasing demand of trained Ayurvedic pharmacy professionals.

At present, there are ten thousand pharmaceutical units producing Ayurvedic medicines in India. These are meeting the domestic and global requirements. The estimated market of these products is about four thousand crore rupees per year.

Ayurveda education at pharmacy level came into limelight after establishment of Institute of Ayurvedic Pharmaceutical Sciences at Gujarat Ayurvedic University (GAU), Jamnagar. The university took the initiative of introducing pharmacy course related to Ayurveda ranging from diploma to doctorate level. National Institute of Pharmaceutical Education and Research (NIPER), Mohali, initiated Masters Program in Traditional Medicine on the persistent demand of the Ayurvedic drug industry in 2008.

13.2. COURSES RELATED TO AYURVEDIC PHARMACY

Courses of study in Ayurvedic pharmacy as offered in different universities are:

Traditional

1. Diploma in Ayurvedic Pharmacy (D. Pharma) – 2 years
2. Bachelor in Ayurvedic Pharmacy (B. Pharma) – 4 years
3. Master in Ayurvedic Pharmacy (M. Pharma) – 2 years.

Interdisciplinary

1. Master of Science in Medicinal Plants (M.Sc. Medicinal Plants) – 2 years
2. PG Diploma in Ayurvedic Drug Standardisation – 1 year
3. MBA (Pharma) specially designed for Pharmacy graduates – 2 years

4. Certificate Course in Ayurvedic Cosmetics – 1 year
5. Certificate Course in Phytochemical Techniques – 6 months.

Diploma in Pharmacy (Ayurveda) – It is abbreviated as D. Pharm (Ayu). The course trains personnel for the consumer sector of Ayurvedic drugs. They will be qualified to dispense Ayurvedic drugs in hospitals, dispensaries, etc., and training shall be imparted for maintaining stores of raw materials & finished products.

Bachelor of Pharmacy (Ayurveda) – It is abbreviated as B. Pharm (Ayu). This course trains individuals to become a good manufacturing pharmacist of classical Ayurvedic formulations, a good quality control pharmacist & a pharmaceutical technologist. The candidate can also enter into the services of food and drug control administration of the state.

13.3. THE INDIAN MEDICINE & HOMEOPATHY PHARMACY COUNCIL BILL, 2005

The Indian Medicine and Homeopathy Pharmacy Council Bill was introduced 2005 in the Parliament. The bill was aimed at regulating education and practice of pharmacists in Ayurveda, Unani, Siddha and Homoeopathy (AYUSH), as there was no standardisation and uniform education system of curriculum available for training pharmacists in this sector. The standing committee attached to the health ministry examined and submitted the report on July 2006, but it has been pending ever since.

Finally the bill was revised by the AYUSH department, as per the recommendations of the committee, and was sent to the Ministry of Law and Justice for vetting. The bill was cleared by the law ministry and was ready for submission to the cabinet for final clearance, but the new proposal of the National Commission for Higher Education and Research (NCHER) made it totally irrelevant.

With the government firm on its move to establish the NCHER, the much-awaited bill to set up a different pharmacy council has gone redundant. Though a revised bill, the Indian Medicine and Homoeopathy Pharmacy Council to replace the original bill of 2005, is pending for Cabinet approval, the Department of AYUSH has completely dropped it in view of the NCHER which will also cover the Ayush courses.

13.4. RECOGNITION ISSUES RELATED TO AYURVEDIC PHARMACY

The National Institute of Ayurveda is the apex institute for training and research in Ayurveda in India for imparting Diploma in AYUSH Nursing and Pharmacy (DAN & P). Department of Ayurvedic Medical Education, Kerala, is offering Ayurvedic pharmacy course under paramedical courses. Department of Ayurveda, Government of Himachal Pradesh has created Ayurvedic Pharmacy College at Jogindernagar District, Mandi.

Several universities in India are imparting courses related to Ayurvedic pharmacy. These courses may be recognised by the University Grants Commission (UGC) but there is no apex body like Pharmacy Council of India (PCI) responsible for recognition of Ayurvedic pharmacy courses being imparted across India. Bachelor of Ayurveda in Pharmacy is included in the degrees specified by the UGC under section 22 of the UGC Act.

Courses like Bachelor of Ayurvedic Medicine and Surgery (B.A.M.S.), Ayurveda Vachaspati (M.D. Ayurveda) and Ayurveda Dhanwantri (M.S. Ayurveda) are recognised by the Central Council of Indian Medicine (C.C.I.M.), Dept. of AYUSH. Until The Indian Medicine and Homoeopathy Pharmacy Council Bill, 2005 is implemented, Ministry of Health and Family Welfare need to take concrete steps to stop the unrecognised courses being run by several institutions across India.

13.5. REGISTRATION ISSUES OF AYURVEDIC PHARMACISTS

In Sri Lanka Consolidated Acts, Ayurveda Act (No. 31 of 1961) – Section 18 states that the Ayurvedic Medical Council is to be the authority responsible for the registration of Ayurvedic practitioners, Ayurvedic pharmacists and Ayurvedic nurses and the regulation and control of their professional conduct.

In India, some State Faculties of Ayurveda and Unani have initiated the registration of Ayurvedic pharmacists. This is not sufficient to safeguard the future of pass out of courses related to Ayurvedic pharmacy. Ayurvedic pharmacy council on lines parallel with apex bodies like Medical Council of India (MCI), Dental Council of India (DCI), Central Council of Indian Medicine (CCIM), Central Council of Homoeopathy (CCH), Nursing Council of India (NCI) and Pharmacy Council of India (PCI) is mandatory to regulate the registration and practice issues of Ayurvedic pharmacists. Students whom pursued an Ayurvedic pharmacy course from State Faculty of Ayurveda and Unani or Government Ayurvedic College may get employment in the government sector. On the contrary, students having passed the Ayurvedic pharmacy course from a deemed or private university are finding it hard to get recognition and registration in India.

CHAPTER 14
PATENT AND IPR ISSUES OF AYURVEDIC FORMULATIONS

14.1. TRADITIONAL KNOWLEDGE DIGITAL LIBRARY (TKDL)

Since time immemorial, India has possessed a rich traditional knowledge of ways and means practiced to treat diseases afflicting people. This knowledge has generally been passed down by word of mouth from generation to generation. A part of this knowledge has been described in ancient classical and other literature, often inaccessible to the common man and even when accessible rarely understood. Documentation of this existing knowledge, has become imperative to safeguard the sovereignty of this traditional knowledge and to protect it from being misappropriated in the form of patents on non-original innovations, which has been a matter of national concern.

India fought successfully for the revocation of turmeric and basmati patents granted by United States Patent and Trademark Office (USPTO) and neem patent granted by the European Patent Office (EPO). As a sequel to this, in 1999, the Department of Ayurveda, Yoga & Naturopathy, Unani, Siddha and Homoeopathy (AYUSH), erstwhile Department of Indian System of Medicine and Homoeopathy (ISM&H) constituted an interdisciplinary Task Force, for creating an approach paper on establishing a Traditional Knowledge Digital Library (TKDL).

TKDL was initiated in the year 2001. TKDL provides information on traditional knowledge existing in the country, in languages and format understandable by patent examiners at International Patent Offices (IPOs), so as to prevent the grant of wrong patents. TKDL thus acts as a bridge between the traditional knowledge, information existing in local languages and the patent examiners at IPOs.

TKDL is a collaborative project between the Council of Scientific and Industrial Research (CSIR), Ministry of Science and Technology and Department of AYUSH, Ministry of Health and Family Welfare, and is being implemented at CSIR. An interdisciplinary team of Traditional Medicine (Ayurveda, Unani, Siddha and Yoga) experts, patent examiners, IT experts, scientists and technical officers are involved in the creation of TKDL for Indian Systems of Medicine.

The project TKDL involves documentation of the traditional knowledge available in the public domain in the form of existing literature related to Ayurveda, Unani, Siddha and Yoga, in digitized format in five international languages which are English, German, French, Japanese and Spanish. Traditional Knowledge Resource Classification (TKRC), an innovative structured classification system for the purpose of systematic arrangement, dissemination and retrieval has been evolved for about 25,000 subgroups against the few subgroups that was available in earlier versions of the International Patent Classification (IPC), related to medicinal plants, minerals, animal resources, effects and diseases, methods of preparations, mode of administration, etc.

14.2. PROTECTION OF TRADITIONAL KNOWLEDGE BY UTILIZATION OF TKDL

Misappropriation of traditional knowledge and bio-piracy of genetic resources are the issues of great concern for all the developing countries. These issues are being pursued at several multilateral forums, such as the Convention on Biological Diversity, TRIPs Council, World Trade Organisation and World Intellectual Property Organisation. However, so far a 'global framework' for the traditional knowledge protection system has not been established. It is mainly for this reason that Mexico had to fight a legal battle for 10 years to get the patent on Enola bean at the United States Patent & the Trademark Office (USPTO) cancelled in July 2009. Similarly, the cancellation of Monsanto Soybean patent in July 2007 at the European Patent Office (EPO) took 13 years of legal battle.

India is the only country in the world to have set up an institutional mechanism – TKDL, to protect its traditional knowledge and to prevent the grant of wrong patents. A collaborative project between CSIR and Department of AYUSH, Ministry of Health and Family Welfare, TKDL is a maiden Indian effort to help prevent misappropriation of traditional knowledge belonging to India at International Patent Offices. It enables the cancellation/withdrawal of wrong patent applications concerning India's traditional knowledge at zero cost and in a few weeks time. In sharp contrast, in the absence of TKDL, it took 10 years (1995–2005) to get the Neem patent invalidated for antifungal properties at THE EPO.

The genesis of TKDL dates back to the Indian effort on revocation of patent on wound healing properties of Turmeric at the USPTO and anti-fungal properties of Neem at EPO. In 2000, the TKDL expert group estimated that about 2000 wrong patents concerning Indian systems of medicine were being granted every year at an international level, mainly due to the fact that India's traditional medical knowledge existing in languages such as Sanskrit, Hindi, Arabic, Urdu, Tamil, etc. was neither accessible nor comprehensible for the patent examiners at the international patent offices.

The TKDL technology integrates diverse disciplines (Ayurveda, Unani and Siddha), languages (Sanskrit, Arabic, Urdu, Persian and Tamil), modern science and modern medicine. It has created a unique mechanism for overcoming the language and format barriers by scientifically converting and structuring the available information contents of 34 million A4 size pages of the ancient texts in five international languages, namely, English, Japanese, French, German and Spanish.

Through TKDL, a Sanskrit verse can now be read in international languages by an examiner at any International Patent Office on his computer screen enabled by two important features of TKDL: Relevant information technology tools and a novel classification system – Traditional Knowledge Resource Classification (TKRC). Today, through TKDL, India is capable of protecting about 0.226 million medicinal formulations similar to those of Neem and Turmeric. On an average, it takes five to seven years for opposing a granted patent at international level, which may cost 0.2–0.6 million US$. One could only imagine the cost of protecting 0.226 million medicinal formulations in the absence of TKDL!

14.3. FUTURE PROSPECTS OF TKDL

The TKDL already has signed TKDL Access Agreements with countries like USA, Canada, UK, Australia, Japan and Germany. Negotiations are under way to conclude the Access Agreement with Intellectual Property Office of New Zealand and some other countries, sources disclosed. Under the agreement, examiners of the patent office can utilize TKDL for search and examination purposes only and cannot reveal the contents of TKDL to any third party unless it is necessary for the purposes of citation.

The TKDL Access Agreement is unique in nature and has in built safeguards on non-disclosure to protect India's interest against any possible misuse. TKDL is also planning to include additional one lakh formulations from Ayurveda, Unani and Siddha. Another move is to make it available to publically funded research and development institutions for promoting research in the field. As of now, the TKDL has transcribed 150 books, containing 2,83,873 formulations. As many as 75 books from Ayurveda, 10 books from Unani, 50 books of Siddha and 15 books of Yoga were included. 96,781 Ayurveda formulations, 1,63,174 formulations of Unani and 22,264 Siddha formulations are among them.

14.4. INTELLECTUAL PROPERTY RIGHT (IPR)

Intellectual property (IP) is a legal concept which refers to creations of the mind for which exclusive rights are recognized. Under intellectual property law, owners are granted certain exclusive rights to a variety of intangible assets, such as musical, literary, and artistic works; discoveries and inventions; and words, phrases, symbols, and designs. Common types of intellectual property rights include copyright, trademarks, patents, industrial design rights, trade dress, and in some jurisdictions, trade secrets.

Intellectual property rights are customarily divided into two main areas:

    1. Copyright and rights related to copyright
        The rights of authors of literary and artistic works (such as books and other writings, musical compositions, paintings, sculpture, computer programs and

films) are protected by copyright, for a minimum period of 50 years after the death of the author.

Also protected through copyright and related (sometimes referred to as "neighbouring") rights are the rights of performers (e.g., actors, singers and musicians), producers of phonograms (sound recordings) and broadcasting organizations. The main social purpose of protection of copyright and related rights is to encourage and reward creative work.

2. Industrial property
   Industrial property can usefully be divided into two main areas:
   * One area can be characterized as the protection of distinctive signs, in particular trademarks (which distinguish the goods or services of one undertaking from those of other undertakings) and geographical indications (which identify a good as originating in a place where a given characteristic of the good is essentially attributable to its geographical origin).
   The protection of such distinctive signs aims to stimulate and ensure fair competition and to protect consumers, by enabling them to make informed choices between various goods and services. The protection may last indefinitely, provided the sign in question continues to be distinctive.
   * Other types of industrial property are protected primarily to stimulate innovation, design and the creation of technology. In this category fall, inventions (protected by patents), industrial designs and trade secrets. The social purpose is to provide protection for the results of investment in the development of new technology, thus giving the incentive and means to finance research and development activities. A functioning intellectual property regime should also facilitate the transfer of technology in the form of foreign direct investment, joint ventures and licensing. The protection is usually given for a finite term (typically 20 years in the case of patents).

14.5. INTELLECTUAL PROPERTY RIGHTS AND TRADITIONAL KNOWLEDGE

When community members innovate within the traditional knowledge framework, they may use the patent system to protect their innovations. However, traditional knowledge as such – knowledge that has ancient roots and is often informal and oral – is not protected by conventional intellectual property systems. This has prompted some countries to develop their own sui generis (specific, special) systems for protecting traditional knowledge.

There are also many initiatives underway to document traditional knowledge. In most cases the motive is to preserve or disseminate it, or to use it, for example, in environmental management, rather than for the purpose of legal protection. There are nevertheless concerns that if documentation makes traditional knowledge more widely available to the general public, especially if it can be accessed on the Internet, this could lead to misappropriation and use in ways that were not anticipated or intended by traditional knowledge holders. At the same time, documentation can help protect traditional knowledge, for example, by providing a confidential or secret record of traditional knowledge reserved for the relevant community only.

14.6. INTELLECTUAL PROPERTY RIGHTS AND AYURVEDA

Ayurveda is getting its due recognition as a rationale system of medicine worldwide despite the fact that the medical and scientific fraternity of the globe have a very strong opposite opinion regarding the safety and efficacy of Ayurvedic medicines. Meanwhile, provisions of Intellectual Property Rights under World Intellectual Property Organization (WIPO) and Patents have attracted many individuals and organizations to explore possibilities of commercial benefits with Ayurvedic traditional knowledge. Although the rules are not favouring the grant of a patent on prior published knowledge, bio piracy managed grant of patent on knowledge of Ayurvedic medicinal plants has been successfully checked with references to a data base from the Traditional Knowledge Digital Library (TKDL).

Current provisions of the patent law of India are obstructive in nature for getting a patent on Ayurvedic medicines. If we have to invite researchers from basic science to ensure quality, safety and efficacy of Ayurvedic medicines, there is an urgent need to amend the laws of patenting with pragmatic promotional policies. This will encourage more patents on numerous pharmaceutical, nutraceutical and cosmaceutical products based on Ayurveda.

CHAPTER 15
AYURVEDIC PHARMACOEPIDEMIOLOGY

Pharmacoepidemiology is a new field developed by the synergy of the fields of clinical pharmacology and epidemiology. In India, both these fields have not flowered as robust disciplines in the medical colleges, industry and government. As a consequence, the endeavour in Pharmacoepidemiology has been infrequent and at a few centers. But with the growing global demand of Ayurveda and its widespread use in India (70% of people), an emergent discipline of Ayurvedic Pharmacoepidemiology is likely to strike deep roots, with interesting avenues for research, education and services.

The Council of Scientific and Industrial Research (C.S.I.R.) has embarked on a major project – the New Millennium Initiative for Technological Leadership of India (NMITLI). NMITLI has diverse components. But a major component is to develop globally acceptable herbal drugs from the Ayurvedic Therapeutic heritage. Three projects have already been initiated; in diabetes mellitus, osteoarthritis and hepatitis.

Ayurvedic Pharmacoepidemiology will be the study of the usage, acceptability, efficacy, safety, complementarity, and cost-effectiveness of Ayurvedic drugs in a large number of people. It will also encompass fields such as Ayurvedic prescription audits, Ayurvedic drugs outlets/utilization, population pharmacodynamics/kinetics and documentation of untoward or unexpected beneficial effects of Ayurvedic drugs. The data collection, wetting, storage and analysis will utilize state-of-the-art software and automation.

The spin-offs and potential contributions of Ayurvedic Pharmacoepidemiology are expected to be sizeable:

1) Usage safety records for Ayurvedic drugs,
2) Data useful for herbal drug registration,
3) Adverse drug reaction registry for rational therapeutic precautions,
4) Drug dosage adjustments in special age or disease groups of patients,
5) Pharmacoeconomics of Ayurveda vis-à-vis marketing and rational drug policy,
6) Discovery of novel beneficial effects as leads for further research,

7) 'Quality of life' (QOL) studies with Rasayana Dravyas,
8) Patterns of drug use across the systems of medicine,
9) Drug interactions likely due to concomitant administration of intersystem drugs and
10) Fulfillment of ethical and cultural obligations for the heritage of healing.

CHAPTER 16
VOLUNTARY CERTIFICATION SCHEME FOR AYUSH PRODUCTS

16.1. INTRODUCTION

The AYUSH products are regulated under the Drugs and Cosmetics Act, 1940 by the Drugs Controller General of India through the State Governments.

The Department of AYUSH has been exploring the possibility of introducing a voluntary product certification scheme for selected AYUSH products to enhance consumer confidence. The matter was discussed in a series of meetings taken by the Secretary (AYUSH) beginning 24 December 2008 and the Quality Council of India (QCI) offered to develop a concept paper on the subject.

On approval of the concept, the Department of AYUSH signed an agreement with the QCI on 27 July 2009 to design the Scheme with **Department of AYUSH being the Scheme owner and QCI being responsible for managing the Scheme**.

The draft Scheme was given to the Department of AYUSH on 3 August 2009 and simultaneously placed on the websites of Department of AYUSH and QCI for public consultation.

The Scheme will be overseen by a **Multistakeholder Steering Committee (MSC)** chaired by the Secretary (AYUSH) with secretariat in QCI. The MSC will be supported by a **Technical Committee** and a **Certification Committee** constituted by QCI.

The Scheme is based on criteria for certification. It has two levels:

A. *Ayush Standard Mark* which is based on compliance to the domestic regulatory requirements.
B. *Ayush Premium Mark* which is based on GMP requirements based on WHO Guidelines and product requirements with flexibility to certify against any overseas regulation provided these are stricter than the former criteria.

Under this scheme, each manufacturing unit would obtain a certification from an approved certification body (CB) which is accredited to appropriate international standards by the **National Accreditation Board for Certification Bodies (NABCB)** and will be under regular surveillance of the certification body.

16.2. GOVERNING STRUCTURE

The governing structure of the Voluntary Certification Scheme for AYUSH Products shall have a Multistakeholder Steering Committee (MSC) at the apex level supported by a Technical and a Certification Committee each with its secretariat in the Quality Council of India (referred to as QCI hereinafter).

A) Representation of a balance of interests such that no single interest predominates.
B) Key interests to include: representatives of AYUSH industry associations, Certification Bodies, Testing Laboratories, Accreditation bodies, representatives of regulatory bodies or other governmental agencies, Academic/Research Bodies and representatives of non-governmental organizations.
C) Offer of membership to individual experts shall be made with great caution and only when a suitable person is not forthcoming as a representative of an organization.
D) Except when a member is appointed in his personal capacity, a person vacates his membership on leaving his organization and a fresh nomination is sought from the member organization.
E) The member organizations shall nominate a principal and an alternate representative on the committee(s).
F) All committees shall be reconstituted every two years to provide representation to organizations like State Governments, Certification Bodies, etc. by rotation, where necessary.

16.3. MULTISTAKEHOLDER STEERING COMMITTEE (MSC)

The MSC shall comprise of the following;

A) Secretary, Department of AYUSH – Chairperson;
B) Central Government Ministries – One Representative each from the concerned Ministries – Department of Commerce, Department of Consumer Affairs, Department of Health, Department of Forests, MOEF, and Department of AYUSH;
C) Regulatory Bodies – Drugs Controller General of India;
D) Accreditation Bodies – One Representative each from NABCB and NABL;
E) Industry Bodies – One industry body of each stream of AYUSH – Ayurveda, Unani, Siddha and Homeopathy;
F) State Governments – 3 (by rotation for 2 years);
G) Testing Laboratories – 3 – One Representative each from PLIM, HPL, and IIIM, Jammu;
H) Certification Bodies – One representative from QCI accredited bodies (2 years by rotation);
I) Research bodies – 2 (2 years by rotation);

J) Chairman APC;
K) NMPB (2 years by rotation);
L) Professional Bodies of practitioners of each discipline – One Representative from each stream of Ayush Practitioners;
M) Representative of Drug Retail Trade Association;
N) Secretary General, QCI;
O) Any other Technical expert(s) as invites for specific meetings, as identified by the Secretariat;
P) Secretariat – Quality Council of India.

16.4. TECHNICAL COMMITTEE (TC)

A) Chairperson
B) Central Council for Research in Ayurveda and Siddha
C) Certification Body – One from those represented on the MSC
D) Pharmacopeial Laboratory for Indian Medicines (PLIM)
E) Industry Bodies – One of each stream of AYUSH
F) Pharmexcil Export Promotion Council – One Representative
G) Department of Ayush – One Representative
H) DCGI – One Representative
I) Accreditation Bodies – One Representative each from NABCB and NABL
J) Research body from MSC – One
K) Secretariat – Quality Council of India

16.5. CERTIFICATION COMMITTEE (CC)

A) Chairperson
B) Department of AYUSH – One Representative
C) Certification Body – 2 CBs by rotation
D) Accreditation Boards – One Representative each from NABCB and NABL
E) DCGI
F) EIC
G) Industry – One of each stream of AYUSH
H) Pharmexcil
I) PLIM
J) Secretariat – Quality Council of India

16.6. ROLES OF ORGANIZATIONS

Department of AYUSH shall be the Scheme Owner and own the Certification Mark(s). It shall establish the MSC while the management of the Scheme shall be the responsibility of QCI.

Quality Council of India (QCI) shall provide the secretariat to the Scheme and the MSC. It shall set up the Technical and Certification Committees and provide secretariat to them. It shall also manage the Scheme on behalf of Department of AYUSH.

National Accreditation Board for Certification Bodies (NABCB) shall be responsible for accrediting certification bodies desirous of participating in the Scheme to appropriate international standards.

National Accreditation Board for Testing and Calibration Laboratories (NABL) shall be responsible for accrediting testing and calibration laboratories to appropriate international standards to support the Scheme.

16.7. COMPLAINTS

16.7.1. The entire system has provisions for entertaining complaints from any stakeholder against any component of the Scheme – the manufacturing units certified under the Scheme, the Certification Bodies approved under the Scheme, the laboratories utilized under the Scheme, and the accreditation bodies, NABCB/NABL, are all required to have a complaints system in place as per standards applicable to them. Anyone having a complaint is encouraged to utilise the available mechanisms.

16.7.2. Any complaint received directly by the Department of AYUSH shall be referred to QCI who in turn will make a reference to the appropriate body against which the complaint is made and monitor it till it is decided upon.

16.7.3. Any complaint received by QCI shall be similarly handled.

16.7.4. A statement on complaints as received above with their status shall be reported to the MSC in each meeting.

16.8. APPEALS

16.8.1. There are provisions for entertaining appeals from the manufacturing units certified/desirous of certification under the Scheme, the Certification Bodies approved under the Scheme, and the laboratories utilized under the Scheme, which shall invariably be utilized.

16.8.2. In case any one aggrieved by the decision of the TC/CC appeals, it shall be handled by the MSC.

16.8.3. In case any one is aggrieved by the decision of MSC appeals, the Chairperson, MSC shall appoint an independent appeals panel to look into the appeal and recommend action to him/her.

16.8.4. In handling appeals, the broad principle that the appeal is handled independently of the personnel involved in the decision appealed against shall be maintained.

16.8.5. A statement of appeals received by the Department of AYUSH/QCI shall be placed before the MSC in each meeting.

16.9. CERTIFICATION CRITERIA

The certification is available at two levels:

A) AYUSH Standard Mark which is based on compliance to the domestic regulatory requirements; and
B) Ayush Premium Mark which is based on either or both of the following options.

Option A: Compliance to GMP Requirements based on WHO Guidelines and levels of contaminants as given in the Certification Criteria document.

Note
i. The requirements of heavy metals shall not be applicable to Ayush products having raw materials of metallic origin provided they are intended for domestic market.
ii. For the time being this certification is available for Herbal products only.

Option B: Compliance to regulatory requirements

16.9.1. For any manufacturer to qualify for AYUSH Premium Mark certification, compliance to the domestic regulation is a prerequisite.

16.9.2. The various criteria mentioned above are as follows:

A) Domestic regulation means regulatory requirements prescribed under the Drugs and Cosmetics Act, 1940 for AYUSH products;
B) GMP Requirements based on WHO Guidelines for AYUSH Premium Mark;
C) Permissible levels of contaminants for AYUSH Premium Mark;
D) Permissible levels of contaminants for AYUSH Standard Mark;
E) Regulations of importing countries – to be identified by the organization seeking certification and provided to the Certification Body.

16.10. OBTAINING AND MAINTAINING CERTIFICATION

16.10.1. STEP I Planning for Product Certification

1. Obtaining certification against the Certification Criteria represents a challenge to manufacturers of AYUSH products.
2. It is therefore essential that the organizations interested in obtaining this certification consider carefully
   a) What Ayush Certification they wish to achieve,
   b) Identify the Criteria and
   c) What needs to be achieved prior to applying for product certification.

16.10.2. STEP II Preparation

1. Obtain the relevant Certification Criteria documents, namely the Domestic Regulations under the Drugs and Cosmetics Act, GMP Requirements based on

WHO Guidelines, Levels of Contaminants, and Regulations of the importing country (if applicable).
2. Assess your process and product for compliance to relevant Certification Criteria.
3. Undertake preparations so as to ensure that the processes and the Ayush product being manufactured would comply with the requirements of the relevant Certification Criteria.
4. Install inhouse testing facilities if seeking certification for Ayush Premium Mark.
5. Confirm that your production facility has been in production for at least one year.
6. Verify that five commercial batches of the products of dosage form for which certification is being planned to be sought have been manufactured during the current licensed period.

16.10.3. Self Assessment

1. Review your current systems and practices against the requirements of the latest relevant Certification Criteria.
2. Identify areas which need to be addressed and ascertain compliance prior to applying for product certification.
3. Please note that Certification Bodies are not permitted to provide consultancy although they can identify deviations from the relevant Certification Criteria requiring correction and/or corrective action.

16.10.4. Select a Certification Body

1. Select an approved Product Certification Body to carry out the evaluation for Product certification at your site. Only Product Certification Bodies that are accredited by NABCB/recommended by QCI can certify for Ayush Product certification.
2. In selecting a certification body, consider the range of products covered under the scope of accreditation of the Product Certification Body. Information on scope of accreditation is available on the websites of Department of AYUSH (www.indianmedicine.nic.in) and QCI (www.qcin.org) or can be obtained from the concerned certification body, either through correspondence or by visiting their website.
3. Certification Bodies will require details of your site, operations, products and relevant certification criteria on a prescribed application form for registration of your application for grant of product certification. Obtain the prescribed application form from the selected Certification body.
4. Submit the application form duly filled to the selected certification body along with the following documents;
   A) Proof of being a legal entity;
   B) Valid Manufacturing Licence;
   C) Proof of Address of production site;
   D) List of installed manufacturing equipment;
   E) List of testing facilities, if available;

F) Copy of a test report covering all requirements of the relevant certification criteria for the product applied for; test report could be either from own laboratory or from an NABL accredited external laboratory;
G) Self evaluation checklist confirming that all requirements as prescribed in the relevant certification criteria are being complied with.

16.10.5. Registration and Evaluation by a Certification Body

1. The certification body will examine your application for completeness and adequacy, and deviations if any, will be informed to you for necessary correction and/or compliance. Once the application is found to be complete, it shall be registered by the Certification body.
2. The certification body will determine the duration of the evaluation at your site, on the basis of the number of products to be certified, the manpower, the complexity of operations and processes, and availability of lab and quote to you.
3. On acceptance of its quote, the Certification body shall in consultation with you finalize the dates for evaluation.
4. It is important that the facility is in production at the time of the Evaluation by the certification body otherwise another evaluation may have to be organized.
5. The certification body will carry out the evaluation in one stage for AYUSH Standard and in two stages for AYUSH Premium Mark. During the Stage 1 evaluation, your state of preparedness, status of GMPs and availability of competent personnel and equipment for production and testing will be assessed for their adequacy.
6. At the end of Stage 1 evaluation the Certification Body will inform the applicant in writing about the deficiencies observed, if any, with respect to the certification criteria.
7. Take necessary actions and inform the certification body as soon as possible but not later than three months of the Stage 1 Evaluation. Delays beyond this will lead to another Stage 1 evaluation of your facility by the certification body.
8. The certification Body will undertake the Stage 2 evaluation only after you have confirmed that necessary actions on the identified shortfalls have been taken.
9. For the Stage 2 evaluation the certification body will visit the facility and evaluate the process and controls being implemented, the prevailing hygienic conditions, the testing facilities and the competence of the personnel for compliance to the certification criteria.
10. The certification body will draw samples of products from stocks that are representative of normal production capacities of the facility, and have the same tested in an independent laboratory accredited by NABL, for compliance to the certification criteria.

16.10.6. Follow Up and Corrective Actions

1. At the end of Stage 2 evaluation the Certification Body will inform the applicant in writing of the deviations observed, if any, with respect to the certification criteria.

2. Take necessary actions and inform the certification body as soon as possible. If you do not show progress towards completion of corrective actions within three months of Initial Evaluation your application shall be closed.
3. If the corrective actions can be verified through documented evidence, you will be required to provide the same to the Certification Body for verification of corrective actions. However if verification of corrective action is to be undertaken at the manufacturing site or if a follow up evaluation is required is only possible at site, the Certification body evaluator visits the site for confirmation of the identified corrective action only or for follow up of non conformities identified during the previous evaluation.
4. If the sample drawn fails on independent testing, you will be informed of the same and advised to take necessary corrections for improving the product quality.
5. After taking the corrective actions, you must re-offer the products for sampling and testing by the certification body. The Certification body plans another visit to the site, verifies the corrective actions taken by the organization and draws the fresh sample from a stock of material that is representative of the normal production capacity of the organization.

16.10.7. Certification Decision

1. The Certification Body will review the onsite Stage 1 (if applicable) and Stage 2 evaluation reports, corrective action documentation provided by the applicant and verified if so required by the Evaluation team, and the independent test report(s) in order to make a certification decision.
2. Certificate shall be awarded to the manufacturer only for the range of products applied for and offered for evaluation and testing.
3. The decision for certification will be taken only when all requirements of the Scheme have been complied with.
4. The Certificate shall be awarded for fixed time tenure of three years, during which your operations and the products will be subjected to surveillance evaluations and testing, and beyond the three years period of validity the certificate will be renewed subject to ongoing compliance.
5. The certification process should be completed within 12 months of the registration of the application failing which the certification body would reject your application.

16.10.8. Issue of a Certificate by Certification Body

1. The Certification Body issues a certificate to the manufacturer indicating that the requirements of the certification scheme have been met with and that the products conform to the relevant certification criteria. The name of the manufacturer with address of site, names of products, and the relevant certification criteria are clearly mentioned on the Certificate, along with effective date of certificate, validity of the certificate, name and address of the Certification Body and applicable logos.

2. The certificate should be issued within seven days of the certification decision.
3. The Certification Body immediately informs the QCI about the grant of product certification.

16.10.9. Agreement for Usage of AYUSH Certification Mark(s)

1. The AYUSH Certification Mark(s) is not available for download from the AYUSH website/any website.
2. Apply for authorization for affixing the applicable Certification Mark on AYUSH products for which you have been certified by the Certification body, on a prescribed format, which can be downloaded from the QCI website.
3. Submit the application form duly filled to the QCI secretariat.
4. Enter into a legally enforceable agreement with QCI authorizing you to affix the Certification Mark on the products for which you have been certified.
5. Based on this, the QCI website is updated with the name and address of the Ayush manufacturer, the product covered in the scope of certification, the criteria against which certified, the Certification mark awarded, effective date of certificate and its validity.

16.10.10. Maintaining Certification

1. Ensure that production and testing facilities are maintained in working order and technical personnel are available at all times.
2. Implement Internal Quality Assurance Protocol provided by the CB and maintain records.
3. Inform production schedules to the CB when asked for.
4. Ensure that the products certified under the Scheme comply with the Certification criteria – this will be checked by the CB through factory evaluation and market surveillance.

16.11. USING THE AYUSH CERTIFICATION MARK(S)

1. The AYUSH Certification Mark(s) is for use only by organisations that have achieved product certification.
2. The AYUSH Certification Mark(s) is to be affixed on the product and its packaging to depict product conformity to requirements of the AYUSH certification criteria.
3. The AYUSH Certification Mark(s) shall be affixed only on products conforming to relevant certification criteria, and non conforming products shall not to be marked with AYUSH Certification Mark(s).
4. The AYUSH Certification Mark(s) is owned by the Department of AYUSH and provided on its behalf to the certified unit. The AYUSH Certification Mark(s) can be used on all the manufacturing unit's communication tools such as company vehicles, letterheads, compliment slips, business cards, marketing collateral, advertising, exhibition graphics and electronic media.

5. Misuse of the AYUSH Certification Mark(s) would invite actions including rejection of application or suspension/cancellation of certification.

16.12. APPROVAL FOR USE OF CERTIFICATION MARKS TO CERTIFIED UNITS

1. This document describes the process that manufacturing units certified under the Scheme are expected to meet for obtaining approval for the use of relevant AYUSH Certification Mark.
2. The AYUSH Certification Mark(s), AYUSH Standard Mark and AYUSH Premium Mark are owned by the Department of AYUSH for depicting the product conformity to the relevant certification criteria.
3. Manufacturing units that have been certified under the Scheme are eligible to apply for approval for use of the Certification Mark(s) on the AYUSH products being manufactured by them.
4. The Department of AYUSH does not levy any fee for approval for use of Certification Mark to Certified Manufacturing Units.
5. The form for Approval for Use of Certification Mark seeking information on name and address of the manufacturing unit, the products covered under the Certificate, the relevant certification criteria, the Certificate number and the validity period, can be downloaded from the website of the QCI.
6. The Certified manufacturing units shall submit the form duly filled in along with a copy of the Certificate or letter informing grant of certification by the approved CB under the Scheme to QCI.
7. The relevant Certification Mark for the certified manufacturing unit is determined by QCI on the basis of the Certification Criteria mentioned on the Certificate issued by the CB.
8. The certified manufacturing unit signs a legally enforceable agreement with QCI whereby it is allowed to use the Certification mark agreeing to:
   a) Use only the Certification mark specified by the Department of AYUSH;
   b) Apply the Certification Mark only to products covered under the product certification;
   c) Apply the Certification Mark on AYUSH products conforming to the certification criteria;
   d) Apply it as per the design and size in the colour scheme prescribed;
   e) Not to make any misleading claims with respect to the Certification mark;
   f) Inform the certification body, about the quantum of production covered under the Ayush Certification Mark(s), at the end of every quarter.
9. QCI maintains and updates the directory of certifications granted under the Scheme posted on its website. The directory carries details of all certificates with their status, name and address of the certified clients, products covered under scope of certification, certification criteria, approval to use AYUSH Standard Mark or AYUSH Premium Mark, and name of the certification body.
10. All certification bodies that operate the Scheme are permitted to use the Ayush Certification Mark(s) on their stationary, media for exchange of any

communication, for promoting the awareness of the scheme, the Certification Mark(s) as well as for informing that they are operating the voluntary certification scheme.
11. The approved product certification body shall not use and permit the use of the Ayush Certification Mark in such a manner as to bring the Department of AYUSH into disrepute.

16.13. ACCREDITATION PROCEDURE

16.13.1. APPLICATION

A. Any certification body interested in the accreditation of NABCB for the purpose of AYUSH products certification may apply to NABCB. New certification body shall send an application using NABCB format BCB: F (PCB) 001. Already accredited CB for product certification shall send a request for scope extension on its letterhead.
B. The filled application form for NABCB accreditation shall be duly signed by the CEO/authorized representative/s of the CB seeking accreditation. Request for extension shall be either e-mailed by authorized person or letter sent shall be duly signed by authorized person.
C. CB shall clearly indicate whether they wish to cover either or both of the following sub scopes:
>Ayurvedic products/medicines
>Homeopathic medicines
>Siddha medicines Unani medicines

16.13.2. ASSESSMENT 2.1: A new CB shall undergo accreditation process as per NABCB accreditation procedure for product certification bodies, BCB 201(PCB). 2.2 In case a CB is already accredited by NABCB for ISO Guide 65 for other scope/s, a one man day document review shall be conducted by NABCB for the documentation relating to VCSAP. 2.2.1 After completion of document review, NABCB will undertake an office assessment which shall generally be for one man day. The technical expert's man-day/s would be charged extra, as applicable. 2.2.2 After successful completion of above assessments, NABCB shall conduct one witness audit for each sub scope applied for under the Scheme.

16.13.3. DECISION: Based on the outcome of the above assessments, the decision on extension of scope/s would be taken.

CHAPTER 17
AYURVEDIC DATABASES

17.1. DATABASE ON MEDICINAL PLANTS

Necessity of developing a comprehensive database on important medicinal plants was realized at National Medicinal Plants Board (NMPB) during 2008. Accordingly a project was sanctioned to Central Council for Research in Ayurvedic Sciences (CCRAS) under Extra Mural Research Projects scheme.

Sixteen important medicinal plants having high trade value were selected in the first phase. Exhaustive references and literature were collected with respect to these plants, covering every subject area. For this, various libraries located at New Delhi and other cities in India and websites were consulted. The reprints and literatures were downloaded/photocopied from the websites and libraries with the permission of respective authorities, sometimes by paying the required amount to do so.

Nearly 33,700 references have been collected pertaining to the 16 selected medicinal plants. Of these, 22,000 reprints/abstracts could so far been collected. These include the classical literature from Ayurveda, Unani, Siddha and Homeopathy system of medicine as well as modern literature from various books and journals covering basic and applied science and medicine.

17.2. AYUSH RESEARCH PORTAL

AYUSH Research Portal is meant for disseminating the knowledge of evidence based research data of AYUSH Systems and the current research at the global level meant for academic purpose. AYUSH research portal contains the research data of Ayurveda, Yoga and Naturopathy, Unani, Siddha, Homoeopathy along with related research done by the allied sciences. It has mainly two applications.

The research articles are on the following categories:

1. Clinical Research (A, B, C gradation was also given to the clinical research articles based on the WHO grading recommendations)
2. Pre-Clinical Research
3. Drug Research

The research portal shows the number of studies/data available in each category. The user can search the data by advanced search options viz., Simple Search, Search with AYUSH Terminology, Search in Article Title, Search in Author's Name, Search in Journal Name and Search in Institute/Department.

17.3. DIGITAL HELPLINE FOR AYURVEDA RESEARCH ARTICLES (DHARA)

DHARA is an online index of articles on Ayurveda published in research journals worldwide. The DHARA project is being implemented by AVT Institute for Advanced Research, the research wing of the Ayurvedic Trust, AVP Group of Institutions, Coimbatore, Tamil Nadu, India. The first phase of the DHARA project was funded by Central Council for Research in Ayurvedic Sciences, Department of AYUSH, Ministry of Health and Family Welfare, Government of India from August 2010 to July 2011.

17.4. NATIONAL LIBRARY OF AYURVEDIC MEDICINE

The NLAM database contains botanical information of about 700 plants which form an integral part of Ayurveda formulations. These herbs can be searched through the SEARCH BOX or can be accessed through individual formulation descriptions.

17.5. AN ANNOTATED BIBLIOGRAPHY OF INDIAN MEDICINE

ABIM started as the bibliography of Jan Meulenbeld's *A History of Indian Medical Literature*, and was first published on the internet as a set of HTML files in 2002.

17.6. This database gives information on Ayurvedic alcoholic beverages and Ayurvedic medicines – *Arishtas* and *Asavas*.

17.7. INTERNATIONAL CATALOGUE OF AYURVEDIC PUBLICATIONS

A project was initiated by the Institute for Post Graduate Teaching and Research in Ayurveda, a constituent institute of Gujarat Ayurved University in collaboration with WHO. In the present DATABASE, database the details of each book entered, is given in a particular order viz. Title of the Book, Name of Author, Total number of Pages, price in India/Foreign currency, edition, Subject, Year of publication, ISBN number, address of publishers and the subject content.

CHAPTER 18

MEDICINAL PLANTS DIVISION OF INDIAN COUNCIL FOR MEDICAL RESEARCH

18.1. INTRODUCTION

The Indian Council of Medical Research (ICMR), New Delhi, the apex body in India for the formulation, coordination and promotion of biomedical research, is one of the oldest medical research bodies in the world. The ICMR is funded by the Government of India through the Department of Health Research, Ministry of Health & Family Welfare.

18.2. RESEARCH PRIORITIES

The Indian Council of Medical Research's research priorities coincide with the National health priorities such as control and management of communicable diseases, fertility control, maternal and child health, control of nutritional disorders, developing alternative strategies for health care delivery, containment within safety limits of environmental and occupational health problems; research on major non-communicable diseases like cancer, cardiovascular diseases, blindness, diabetes and other metabolic and hematological disorders; mental health research and drug research (including medicinal plants and traditional remedies). All these efforts are undertaken with a view to reduce the total burden of disease and to promote health and well-being of the population.

18.3. MEDICINAL PLANTS DIVISION

The global resurgence of interest in medicinal plants has resulted in enormous scientific research/information on multidisciplinary aspects in the past several decades. Scientific research activities, focused not only on various aspects/issues related to medicinal plants considered most promising by researchers, drug industry or in the health care system, but also on thousands of less/un-investigated medicinal plants, mentioned in the ancient literature. However this scientific data is widely scattered.

The focus of the programme is to consolidate the work done in various institutions/ laboratories in India on Indian medicinal plants and presents the multidisciplinary information in the form of Review Monographs.

18.4. REVIEWS ON INDIAN MEDICINAL PLANTS

As part of this programme, the Medicinal Plants Division of the ICMR has brought out thirteen volumes in a series of publications entitled Reviews on Indian Medicinal Plants consolidating multidisciplinary scientific published research work on about 3679 medicinal plants species. Each monograph includes regional names of each medicinal plant, its Sanskrit synonyms as well as the Ayurvedic description (wherever available), ethnobotanical studies, apart from the habitat and the parts used, on one hand, and the details of botanical, pharmacognostical, chemical, pharmacological and clinical data on the other, backed by complete references and bibliography on each aspect of the information cited, besides the colour photographs of important medicinal plants. The publication includes about 56964 citations on various aspects.

18.5. QUALITY STANDARDS OF INDIAN MEDICINAL PLANTS

One of the stumbling blocks in the popularity and wider acceptance of Indian herbal drugs had been the inadequacy or lack of standards. This is primarily attributed to the raw material which is not of desired quality as required for reliable biological, pharmacological, chemical and clinical evaluation apart from their use in health care. It is imperative to have quality parameters laid down for the crude raw material to ensure their quality, safety and efficacy. In view of this, ICMR initiated a program for laying down quality standards of medicinal plants widely used in India.

As an outcome of the ongoing programme carried out, ICMR has brought out 12 volumes of Quality Standards of Indian Medicinal Plants. It contains quality standards for 414 medicinal plants commonly used in India.

These volumes on Quality Standards of Indian Medicinal plants cover authentic botanical names, synonyms, parts used, detailed macroscopic and microscopic description of the plant part used as drug, its diagnostic characters and colour photographs. Other parameters in the just released publication include limits of foreign matter, total ash, acid insoluble ash and ethanol extractive, water soluble extractive besides chromatographic finger print profiling. Information on important pharmacological, clinical studies, toxicity, if any, as reported in the scientific literature and prescribed in the Ayurvedic texts supported by complete references, are some of the other highlights of the monographs. The work on other volumes in the series is under progress.

18.6. PHYTOCHEMICAL REFERENCE STANDARDS OF SELECTED INDIAN MEDICINAL PLANTS

The programme aims at generation of Phytochemical Reference Standards (PRS), for some frequently used Indian medicinal plants. Under this programme, simple procedures of isolation of PRS have been optimized by making appropriate modifications in the reported procedures. The quantification of the PRS in the total crude extract is carried out by HPLC. Special emphasis has been laid on the complete characterization of the purified PRS by recording the physical and spectral data. Extensive use of the modern spectral methods like UV-Vis, IR 1H NMR and 13C NMR spectroscopy have been done for this purpose. A herbarium sheet of each of these plants is also maintained along with the repository. The first volume of Phytochemical Reference Standards of Selected Indian Medicinal Plants containing monographs of 30 marker compounds was published.

18.7. MEDICINAL PLANTS MONOGRAPHS ON DISEASES OF PUBLIC HEALTH IMPORTANCE

Inspite of tremendous advancements in drug research, modern synthetic drugs are still not available or adequately suitable a for variety of diseases, particularly new and reemerging infections and the spread of deadly strains that are resistant to various drugs, for which there is no cure in the allopathic system of medicine and post global health challenge.

In view of the above, an activity has been initiated for preparation of monographs on select diseases of public health importance matching with the national health priorities. The monographs shall incorporate information on diseases (including etiopathogenesis) and plant drugs as given in the ancient texts (Indigenous system of Medicine) and the allopathic system of medicine on one hand and the multidisciplinary research data generated during various scientific studies on these plant remedies with focus on pharmacological, toxicological, clinical, phytochemical, pharmacognostic, on the other. The monographs shall also include complete references of the work cited and the photographs of important medicinal plants.

Bibliography
(Part A)

Anonymous. WHO Traditional Medicine Strategy 2002–2005. Geneva: WHO; 2002.
Anonymus. National Pharmacovigilance Protocol for ASU Drugs, IPGTRA. Gujarat Ayurveda University; 2008.
Arora D. Pharmacovigilance obligations of the pharmaceuticals companies. *Indian J Pharmacol* 2008; 40: 13–6.
Bais S, Chandewa AV. Toxicological standardisation of marketed Ashwagandha formulations by atomic absorption spectroscopy. *Asian J Pharm Clin Res* 2013; 6: 45–8.
Chaudhary A, Singh N, Kumar N. Pharmacovigilance: Boon for the safety and efficacy of Ayurvedic formulations. *J Ayurveda Integr Med* 2010; 1: 251–6.
Dahanukar SA, Thatte UM. Can we prescribe Ayurvedic drugs rationally? *Indian Pract* 1998; 51: 882–6.
De Broe ME. Chinese herbs nephropathy and Balkan endemic nephropathy: Toward a single entity, aristolochic acid nephropathy. *Kidney Int* 2012; 81: 513–5.
Debelle FD, Vanherweghem J-L, Nortier JL. Aristolochic acid nephropathy: A worldwide problem. *Kidney Int* 2008; 74: 158–169.
Ernst E. Heavy metals in traditional Indian remedies. *Eur J Clin Pharmacol* 2002a; 57: 891–6.
Ernst E. Toxic heavy metals and undeclared drugs in Asian herbal medicines. *Trends Pharmacol Sci* 2002b; 23: 136–9.
Galib M, Acharya R. National Pharmacovigilance Programme for Ayurveda, Siddha and Unani Drugs. *AYU* 2008; 29: 191–3.
Garg M, Das S, Singh G. Comparative physicochemical evaluation of a marketed herbomineral formulation: *naga bhasma*. *Indian J Pharm Sci* 2012; 74: 535–40.
Garnier R, Poupon J. Lead poisoning from traditional Indian medicines. *Presse Med* 2006; 35: 1177–80.
Gluhovschi G, Margineanu F, Velciov S, Gluhovschi C, Bob F, Petrica L, Bozdog G, Trandafirescu V, Modalca M. Fifty years of Balkan endemic nephropathy in Romania: some aspects of the endemic focus in the Mehedinti county. *Clin Nephrol* 2011; 75: 34–48.
Gunturu KS, Nagarajan P, McPhedran P, Goodman TR, Hodsdon ME, Strout MP. Ayurvedic herbal medicine and lead poisoning. *J Hematol Oncol* 2011; 20; 4: 51.
Heinrich M, Chan J, Wanke S, Neinhuis C, Simmonds MS. Local uses of Aristolochia species and content of nephrotoxic aristolochic acid 1 and 2—a global assessment based on bibliographic sources. *J Ethnopharmacol* 2009; 125: 108–44.
Ioset JR, Raoelison GE, Hostettmann K. Detection of aristolochic acid in Chinese phytomedicines and dietary supplements used as slimming regimens. *Food Chem Toxicol* 2003; 41: 29–36.
Kales SN, Saper RB. Ayurvedic lead poisoning: An under-recognized, international problem. *Indian J Med Sci* 2009; 63: 379–81.
Kales SN, Christophi CA, Saper RB. Hematopoietic toxicity from lead-containing Ayurvedic medications. *Med Sci Monit* 2007; 13: 295–8.
Khandpur S, Malhotra AK, Bhatia V, Gupta S, Sharma VK, Mishra R, Arora NK. Chronic arsenic toxicity from Ayurvedic medicines. *Int J Dermatol* 2008; 47: 618–21.
Lai MN, Wang SM, Chen PC, Chen YY, Wang JD. Population-based case-control study of Chinese herbal products containing aristolochic acid and urinary tract cancer risk. *J Nat Cancer Inst* 2009; 102: 179.
Lavekar GS, Ravishankar B, Rao SV, Gaidhani SN, Ashok BK, Shukla VJ. Safety/toxicity studies of ayurvedic formulation-*Navratna rasa*. *Toxicol Int* 2009; 16: 37–42.

Lavekar GS, Ravishankar B, Gaidhani S, Shukla VJ, Ashok BK, Padhi MM. *Mahayograj guggulu*: Heavy metal estimation and safety studies. *Int J Ayurveda Res* 2010; 1: 150–8.

Mathur A, Sankar V. Standards of reporting Ayurvedic clinical trials—is there a need? *J Ayurveda Integr Med* 2010; 1: 52–5.

Michl J, Jennings HM, Kite GC, Ingrouille MJ, Simmonds MS, Heinrich M. Is aristolochic acid nephropathy a widespread problem in developing countries? A case study of *Aristolochia indica* L., in Bangladesh using an ethnobotanical-phytochemical approach. *J Ethnopharmacol* 2013; 26; 149: 235–44.

Mishra S. *In vitro* study on heavy metal containing Ayurvedic drug induced genotoxicity in human lymphocyte. *J Global Biosci* 2014; 3: 614–8.

Muzi G, Dell'Omo M, Murgia N, Curina A, Ciabatta S, Abbritti G. Lead poisoning caused by Indian ethnic remedies in Italy. *Med Lav* 2005; 96: 126–33.

Nagarajan S, Sivaji K, Krishnaswamy S, Pemiah B, Rajan KS, Krishnan UM, Sethuraman S. Safety and toxicity issues associated with lead-based traditional herbo-metallic preparations. *J Ethnopharmacol* 2014; 151: 1–11.

Narang AP. Arsenicosis in India. *J Toxicol Clin Toxicol* 1987; 25: 287–95.

NTP (National toxicological programme), United states department of health and human services research triangle park, NC 27709.2008. Final report on carcinogens background document for aristolochic acids; 1–12.

Panda AK, Hazra J. Arsenical compounds in ayurveda medicine: A prospective analysis. *IJRAP* 2012; 3: 772–6.

Pipasha B. Setting standards for proactive pharmacovigilance in India: The way forward. *Indian J Pharmacol* 2007; 39: 124–8.

Prakash S, Hernandez GT, Dujaili I, Bhalla V. Lead poisoning from an Ayurvedic herbal medicine in a patient with chronic kidney disease. *Nat Rev Nephrol* 2009; 5: 297–300.

Prakash, MMS Kinthada, Naidu PVS, Muralidhar P. Biologically estimation of heavy/toxic metals present in traditional medicinal plant – *Eclipta alba. Int J Pharm Biomed Sci* 2011; 2: 99–102.

Rajashekharan S, Pushpangadan P, Kumar PKR, Jawahar CR, Nair CPR, Amma LS. Ethno-medico-botanical studies of cheriya arayan and valiya arayan (*Aristolochia indica*, Linn; *Aristolochia tagala*, Cham.). *Ancient Sci Life* 1989; 9: 99–106.

Saper RB, Kales SN, Paquin J, Burns MJ, Eisenberg DM, Davis RB, Phillips RS. Heavy metal content of ayurvedic herbal medicine products. *JAMA* 2004; 292: 2868–2873.

Saper RB, Phillips RS, Sehgal A, Khouri N, Davis RB, Paquin J et al. Lead, mercury, and arsenic in US and Indian-manufactured Ayurvedic medicines sold over the internet. *JAMA* 2008; 300: 915–23.

Saralla RP, Lavanya R, Sudha V, Brindha P. Quality control studies of *Chandraprabha vati*. *Int J Pharm Pharm Sci* 2013; 6: 80–4.

Sathya T, Murthy B, Vardhini N. Genotoxicity evaluation of certain Bhasmas using Micronucleus and Comet assays. *The Internet J Altern Med* 2008; 7.

Satkopan S. Jamnagar: Gujarat Ayurveda University; 2000. Quality Control and Standardisations of Ayurvedic Medicines, Proceedings of National Workshop for Internationally Acceptable Protocol of Ayurvedic Formulations, IPGTRA.

Shukla A. Pharmacovigilance – A gateway to safety and efficacy of ayurvedic drugs. *Int J Ayu Pharm Chem* 2014; 1: 170–5.

Singh KP, Bhattacharya S, Sharma P. Assessment of heavy metal contents of some Indian medicinal plants. *American-Eurasian J Agric Environ Sci* 2014; 14: 1125–9.

Singh P, Yadav RJ, Pandey A. Utilization of indigenous systems of medicine & homoeopathy in India. *Indian J Med Res* 2005; 122: 137–42.

Singh S, Mukherjee KK, Gill KD, Flora SJ. Lead-induced peripheral neuropathy following Ayurvedic medication. *Indian J Med Sci* 2009; 63: 408–10.

Strom BL. Pharmacoepidemiology. Churchill Livingston, New York, 1989.

Thatte U. Proceedings of Preconference workshop "Pharamacovigilance of Ayurvedic Medicines", 2006.

Thatte U, Bhalerao S. Pharmacovigilance of Ayurvedic medicines in India. *Indian J Pharmacol* 2008; 40: 10–2.

Treleaven J, Meller S, Farmer P, Birchall D, Goldman J, Piller G. Arsenic and Ayurveda. *Leuk Lymphoma* 1993; 10: 343–5.

Vaidya ADB, Vaidya RA, Nagral SI. Ayurveda and different kind of evidence: From Lord MaCaulay to Lord Walton (1835 to 2001 AD). *JAPI* 2001; 49: 534–7.

Vanherweghem JL. Aristolochia sp. and chronic interstitial nephropathies in Indians. *Lancet* 1997; 349: 1399.

Vanherweghem JL, Depierreux M, Tielemans C, Abramowicz D, Dratwa M, Jadoul M et al. Rapidly progressive interstitial renal fibrosis in young women: Association with slimming regimen including Chinese herbs. *Lancet* 1993; 341: 387–91.

Vardhini NV, Sathya TN, Balakrishna MP. Assessment of genotoxic potential of herbomineral preparations-*Bhasmas*. *Curr Sci* 2010; 99: 1096–100.

Wu, Tian-Shung et al. Chemical constituents and pharmacology of Aristolochia species. In Rahman, Atta-ur. Studies in Natural Products Chemistry: Bioactive Natural Products (Part L). Gulf Professional Publishing. 2005; 863.

PART B
PHARMACOLOGICAL INVESTIGATIONS ON AYURVEDIC FORMULATIONS

A-Z OF STANDARDISATION, PRE-CLINICAL, CLINICAL AND TOXICOLOGICAL DATA

Abha-Guggulu

Clinical study

Fractures: A clinical trial was carried out in sample of 50 patients having simple long bone fracture viz. - 35 in trial and 15 in control group. Trial group patients were given *Abha-Guggulu* 1 gm b.d. for 45 days where as control group patients were given only reduction & immobilization. The clinical result & scientific evaluation concluded that the conservative way of treating fracture with *Abha-Guggulu* is highly effective (Tripathy and Panda 2009).

Abhraka Bhasma (Mica Calyx)

Analytical Study: EDXRF revealed the presence of Fe (22%) as a major element and Ca, K and Si in low concentrations, their concentration being 11%, 8% and 13% respectively. Mg (4%), Al (2%) and Ti (1%) were present as minor elements while Sodium, Chlorine, and Phosphorous were present in traces (<1%). FEG-SEM studies showed that the grains in *Abhraka Bhasma* were heterogenous and in aggregates of particle size between 19 nm and 88 nm. The grains were found to be irregular in shape ranging from spherical to oblong (Bhatia and Kale 2013).

Pre-clinical pharmacology

Hypoglycemia: Kadam and co-workers reported hypoglycaemic action of *Abhraka bhasma* in alloxan induced hyperglycemia (Kadam et al. 2003).

Hepatoprotective: Different doses of *Abhraka bhasma* (10, 20, 30 and 40 mg/kg body wt) were tested to decide the dose related hepatoprotective efficacy in albino rats using a model of hepatitis induced by a single dose of CCl4 (3 ml/kg body wt). *Abhraka bhasma* counteracted the action of CCl4 on liver lipolytic enzymes. CCl4 did not alter the kidney histologically. Activities of three lipases of rat kidney (acid, alkaline and lipoprotein lipases) were reduced by CCl4 treatment and were reversed by administration of *abhraka bhasma* (Buwa et al. 2001).

To compare their hepatoprotective potency, *Abhraka bhasma* and Silicon dioxide were tested against the single dose of CCl4 induced hepatotoxicity in a male albino rat. Graded single doses viz. 10 mg, 20 mg, 30 mg and 40 mg of *Abhraka bhasma* and SiO2/kg body wt were used simultaneously. *Abhraka bhasma* counteracted CCl4 induced MDA levels to bring them to normal levels. In kidney, CCl4 induced levels of MDA were low which were protected by all the doses of *abhrak bhasma* and high doses of SiO2, i.e., 30 and 40 mg/kg body wt (Teli et al. 2014).

Anti-impotency: The current experiment was carried out on 32 healthy adult male albino Wistar rats divided into four groups. *Sahastraputi Abhraka Bhasma*, subjected to 1000 putas, was used as the test drug. On sacrificing the animals after 30 days, it was observed that control animals (G1) had normal spermatogenesis and drug-induced animals (G2) showed hyperactive tubules. Testicular hyperthermia occurred in few (G3) animals, who were subjected to 43°C for 1 h daily for four consecutive weeks, resulting in degeneration of tubules with inspissated spermatozoa (25%) leading to atrophy of the organ. 3% tubules showed disintegration, 23% were in the recovery stage while 71% tubules exhibited enhanced proliferation of germinal epithelium leading to hypertrophy and hyperplasia (Bhatia et al. 2012).

Abhayarishta

Standardization: Standardisation procedure of the medicinal plants used in *Abhayarishta* has been done (Sharma and Jolly 1992). A simple, rapid, precise and accurate gas liquid chromatographic method has been developed for determination of ethanol in *abhayarishta* (Wadher et al. 2007). An HPLC-DAD method for quantitative estimation of selected marker constituents in the formulation has been developed and validated. A comparison of decoction and final processed formulation revealed chebulagic and chebulinic acid of *Terminalia chebula* were hydrolyzed to their respective monomers and, consequently, there was an increase in the amount of chebulic acid, gallic acid, ellagic acid and ethyl gallate after fermentation. 5-Hydroxymethyl furfural was also found (Lal et al. 2010).

Agnimukha churna (*Narsingh churna*)

Pre-clinical pharmacology

Gastroprotective: Alcoholic Extract (200 mg/kg) and hydro-alcoholic extract (200 mg/kg) of *Agnimukha churna* maintained the integrity of gastric mucosa by virtue of its effect on offensive and defensive gastric mucosal factors. The extracts significantly ($P < 0.0001$) decreased ulcer index in ethanol induced and aspirin induced model (Rao et al. 2014).

Toxicity: In an acute toxicity study, *Agnimukha churna* was non-toxic up to the dose level of 2000 mg/kg p.o. In a subacute toxicity study, *Agnimukha churna* was tested at different dose levels of 250, 500 and 1000 mg/kg p.o. once daily for 28 days. There were no adverse effects on general condition, growth, feed and water consumption, clinical chemistry values and hematological parameters.

Clinical study

Rheumatoid arthritis: A pilot study reported effect of *Narsingha churna* in the management rheumatoid arthritis (Sharma and Borkar 2003).

Agnitundi Rasa

Clinical study

Chorea: A study reported a case of eleven years old child who was treated with the combination of *Mahayograj Guggulu, Rasraj* and *Agnitundi Rasa* (Verma 1985).

Ajmodadi churna

Standardization: *Ajmodadi churna* was prepared as per Ayurvedic Formulary of India. In-house preparation and the marketed drug have been standardized on the basis of organoleptic characters, physical characteristics and physico-chemical properties (Sriwastava, Shreedhara and Aswatha Ram 2010a). In yet another study, standardization of *Ajmodadi churna* has been done on piperine content (Gupta and Jain 2012).

Preclinical pharmacology

Anti-inflammatory: Aqueous extracts of *Ajmodadi churna* significantly reduced paw edema, during the second phase of edema development. In the carrageenan-induced air pouch model, AJM inhibited cellular infiltration into the air pouch fluid (Aswatha Ram et al. 2012).

Antioxidant: The extracts of *Ajmodadi churna* showed good dose dependent free radical scavenging property in all the models. IC_{50} values of methanolic and aqueous extracts were found to be 98 µg/ml and 100 µg/ml for DPPH, 145 µg/ml and 310 µg/ml for ABTS, 600 µg/ml and 620 µg/ml for nitric oxide scavenging activity. The extract showed significant antioxidant activity in all models when compared with ascorbic acid having IC50 values 11 µg/ml, 14 µg/ml and 220 µg/ml for DPPH, ABTS, and nitric oxide scavenging activity respectively (Sriwastava, Shreedhara and Aswatha Ram 2010b).

Clinical study

Rheumatoid arthritis: Present clinical trial was carried out on 20 patients by using castor oil, dry fomentation and a polyherbal compound *Ajmodadi churna*. After three months of therapy, quite an improvement in symptoms of rheumatoid arthritis was observed.

In a study involving 15 patients, *Ajmodadi churna* proved to be effective in lowering the titre of rheumatoid factor, C-reactive protein, erythrocyte sediment rate, anti-cyclic citrullinated phosphate antibody and symptoms of rheumatoid arthritis (Mishra and Rai 2014).

Alambushadi Ghan Vati

Clinical study

***Rheumatoid arthritis*:** Statistically significant improvement was observed in clinical, functional and hematological parameters in group A (*Alambushadi Ghan Vati* Group) and no improvement was found in these parameters in group B (Placebo group) after the course of treatment (Singh and Tiwari 2010).

Amavatavidhvansa rasa

Pre-clinical pharmacology

***Anti-inflammatory*:** The anti-inflammatory activity of *Amavatavidhvansa rasa* against carrageenan-induced paw edema shows that all the three doses had a significant effect and markedly reduced the swelling of paw (Mandavkar and Jalalpure 2014).

Amalakayas Rasayana

Pre-clinical pharmacology

***Adaptogenic*:** A forced swimming hypothermia pre-treatment with *Amalakayas Rasayana* caused significant attenuation of rectal temperature when compared with both stress control and vehicle control groups. It has shown a significant reduction in ulcer index and lipid peroxidation. Moreover, *Amalakayas Rasayana* did show a significant increase in total glutathione content.

***Antioxidant*:** Superoxide radical scavenging activity was done by ethylene diamine tetra acetate and Nitro Blue Tetrazolium Chloride assays against ascorbic acid and $R2$ was 0.976 (EC50 = 77.5 µg/ml). Ferrous reducing power was evaluated by Oyaizu method where $R2$ was 0.986. All studies showed that *Amalakayas Rasayana* possesses antioxidant activity (Samarakoon, Chandola and Shukla 2011b).

***Immunomodulator*:** *Amalaki Rasayana* possesses significant immunostimulant activity and moderate cytoprotective activity. *Amalaki Rasayana* was found to have better activity profile in terms of both immunostimulant as well as cytoprotective activity (Rajani et al. 2012).

Amalkadi Ghrita

Preclinical pharmacology

***Hepatoprotective*:** Administration of *Amalkadi Ghrita* (100 and 300 mg/kg, p.o.) markedly prevented CCl4-induced elevation of levels of serum GPT, GOT, ACP, ALP, and bilirubin. The decreased level of total proteins due to hepatic damage induced by CCl4 was found to be increased in *Amalkadi Ghrita*-treated group. The results are comparable to that of silymarin (Achliya et al. 2004).

Amlapitta Mishran

Preclinical pharmacology

Gastroprotective: Effect of two different doses of *Amlapitta Mishran* was studied by calculating the total number of ulcers, ulcer index and percentage inhibition. *Amlapitta Mishran* treated rats have shown significant ($P < 0.0001$) decrease in the total number of ulcers and ulcer index and significant increase in percent inhibition of ulcers as compared with the positive control group (Vemula et al. 2012).

Anu Taila

Standardisation: Anu taila was packed in white glass bottles and kept in natural light. After four years, different values have been determined and compared with initial values. After four years, sp. gravity, loss on drying, volatile content, acid, saponification, iodine, ester, unsaponification, degree of splitting and double bond values were found less, while free fatty acids, refractive index, agni and neutral taila content were found higher in anu taila (Saxena et al. 1992).

Clinical study

Migraine: Patients diagnosed with migraine were divided in to three groups G1, G2, and G3 and prescribed *Anu taila nasya, Kaphaketu Rasa vati* orally and both, respectively. The trial was conducted for a period of one month. The results were encouraging and maximum relief was observed in the half sided headache, tinnitus and intolerance to light (Madaan and Sharma 2005).

Apamarga Kshara

Evaluation: Samples of *Apamarga Kshara* were subjected to macroscopic, physic-chemical parameters and Sodium, Potassium and Iron assay. The set parameters were found to be sufficient to evaluate the *Apamarga Kshara* and can be used as reference for the quality control/quality assurance.

Clinical research

Hemorrhoids: A total of 30 patients were treated by local application of *Apamarga kshara*. The *ksharapatan* was done every day, for 7 days in 3 g dose, and the result was assessed thoroughly on the basis of observation according to the specially designed proforma. Patients suffering from hemorrhoids were selected by simple random sampling method, with the complaints of bleeding per rectum, pain, discharge, pruritis and prolapse. *Ksharapatan* showed significant improvement in 1st and 2nd degree of pile masses without any side effect (Dudhamal et al. 2010).

Indolent ulcer: *Apamarga Kshara* preparation can be used as a drug for local application after every day cleaning and dressing of the infected wound with *Apamarga Kshara Tailam* (Apaturkar et al. 2013).

Vitiligo: A clinical study reported efficacy of local application with *Jyotishmati-Apamarga Kshara Taila* in the management of vitiligo in children (Arun Raj et al. 2013).

Apamarga Kshara Tailam

Role of media in preparation: All the samples of *Apamarga Kshara Tailam* were tested through various analytical parameters, that is, pH, acid value, iodine value, saponification value, and soon. Finally, it was found that *Apamarga Kshara Tailam* prepared by using fresh *Kalka* and *Ksharajala* was better and it was also an easy pharmaceutical procedure (Gohil et al. 2010).

Clinical research

Indolent ulcer: *Apamarga Kshara* preparation can be used as a drug for local application after every day cleaning and dressing of the infected wound with *Apamarga Kshara Tailam* (Apaturkar et al. 2013).

Arjunarishta (*Parthadyarishta*)

Standardization: Earlier, a standardisation procedure for *Arjunarishta* was developed, however, a detailed study is missing (Vasanthakumar et al. 2003). A TLC- method has been developed for the standardization of *Arjunarishta* by quantitative estimation of major antioxidant compounds, ellagic acid, as markers. Arjunolic acid and arjunic acid were not detected in the formulation (Himmat et al. 2010).

The HPLC analysis showed an increase in amount of ellagic acid and gallic acid during preparation, i.e., decoction vs. formulation. A similar increase in free radical scavenging activity of formulation vs. decoction was also observed. Arjunolic acid and arjunic acid were not detected in the formulation (Lal et al. 2009). A sensitive, simple and accurate HPTLC method has been developed for the quantification of gallic acid in *Arjunarishta* (Sharma et al. 2011).

Arka lavana

Comparative evaluation: *Arka lavana* was prepared in two different methods viz., Traditional puta method and the Modern muffle furnace method and named as *Arka lavana* – A and *Arka lavana* – B respectively. The formulations thus prepared were analyzed with standard parameters like physicochemical characterization. Arka lavana was characterized using FT-IR spectroscopy, Particle size analyzer, Zeta sizer, TG-DTA and X-ray Fluorescence spectroscopy. Sodium and Chloride content was 52.09%, 22.40% and 49.37%, 22.91% in *Arka lavana* – A and B respectively (Devanathan et al. 2013).

Arkadi Kvatha Churna

Toxicity: *Arkadi Kvatha Churna* was administered for 46 days orally to albino rats of both sexes. Animals were fasted for 18 hours after the last administration of *Arkadi Kvatha Churna*. Serum protein & Albumin contents were significantly ($p < 0.05$) increased. Triglyceride and urea level were decreased significantly ($p < 0.05$). Bilirubin contents were decreased significantly ($p < 0.05$) in male rats (Al-Amin et al. 2013).

Arogyavardhini vati

Chemical Analysis: An effort was made to carry out an analytical study (Patgiri et al. 1999). The results from IR and HPTLC revealed presence of metal oxides and organic matter from plant material, whereas ICP-MS method showed the presence of various proportions of elements such as Na, K, Ca, Cr, Mg, V, Fe, Cu, Mn, Ni, Zn which have been found in mg/g and some in μg/g amounts in addition to the major constituent element (Avula et al. 2011).

Preclinical pharmacology

Antioxidant: In an antioxidant assay, *Arogyvardhini vati* 10 mg/ml and 20 mg/ml showed the significant reduction of malondialdehyde concentration and significant improvement in glutathione, superoxide dismutase, catalase amylase activity. *Arogyvardhini vati* demonstrated significant anti-oxidant activity compared to standard and control (Sarashetti et al. 2013).

Hypolipidemic: An experimental study investigated anti-hyperlipidemic activity of *Arogyavardhini vati* against Triton WR-1339-induced hyperlipidemia in rats. *Arogyavardhini vati* significantly decreased serum cholesterol, triglyceride, LDL, and C-reactive protein and significantly increased serum HDL in a dose-dependent manner. Decreased MDA and increased GSH levels in liver were observed at all doses of *Arogyavardhini vati* (50, 100, 200 mg/kg) and fenofibrate-treated groups when compared with Triton-treated group. Atherogenic Index level was significantly decreased in fenofibrate and *Arogyavardhini vati* (200 mg/kg) treated rats when compared with normal control (Kumar et al. 2013).

Clinical pharmacology

Dyslipidemia: In the present prospective cohort study, 108 patients were screened at CGHS Ayurvedic Hospital, New Delhi. Ninety-six patients with satisfied inclusion criteria, and signed informed consent and detailed medical history were recorded. *Arjuna* powder (5 g, BD) for 3 weeks and then *Arogyavardhini Vati* (500 mg, BD) for 4 weeks were prescribed to the patients. The male and female patients were 65.5% (57/87) and 34.5% (30/87), respectively. There was a significant reduction in total cholesterol, LDL, triglycerides, CRP, and blood glucose. However, raised HDL level was also observed (Kumar et al. 2012).

Clinical study

Jaundice: A clinical study regarding efficacy of *Arogyavardhini vati* and *Phalatrikadi kashaya* has been published (Dwivedi et al. 1984).

Leucoderma: A study regarding efficacy of concomitant therapy of *Gandhaka Rasayana* with *Arogyavardhini vati* in the treatment of leucoderma has been published (Shetty et al. 2000).

Otorrhoea: *Arogyavardhini vati* with *prakshalan* of *Kshirivriksha* has been found useful in management of otorrhoea (Shukla et al. 2013).

Psoriasis: A study regarding efficacy of concomitant therapy of *Arogyavardhini vati*, *Kaishore guggulu* and *Chakramardakera taila* in the management of psoriasis has been published (Singh et al. 2007).

Safety evaluation: Earlier, several studies have been performed to study toxicity of *Arogyavardhini vati* (Patigiri et al. 2001). *Arogyavardhini vati* at doses of 50, 250 and 500 mg/kg (1, 5 and 10 times of human equivalent dose respectively), mercury chloride (1 mg/kg) and normal saline were administered orally to male Wistar rats for 28 days. Behavioral parameters were assessed on day 1, 7th, 14th and 28th using Morris water maze, passive avoidance, elevated plus maze and rota rod. Mercury chloride treated group as well as *Arogyavardhini vati* treated groups (50, 250 and 500 mg/kg) showed increased levels of mercury in brain, liver and kidney as compared to normal control (Kumar et al. 2012).

Aravindasava

Fermentation studies: *Aravindasava* was prepared as per the textual and modified methods. The modified methods involved the use of glass vessel and inoculation of the autoclaved drug with the yeasts isolated from *Dasamularista* and *Pippaliasava*. The quantity of alcohol produced in the glass vessel was more than that in the earthen pot by classical method. Among the inoculated organisms, *Dasamularista* yeast II showed highest alcohol production (Alam et al. 1986).

Ashokarishta

Physico-chemical study: The study of the effect of Potassium permagnate shows that non-reducing sugar gets oxidized and gives alcohols acids and tannins. Total solid and non-reducing sugars have been decreased. Viscosity of *Ashokarishta* water has been measured at 302 ± 0.01 k (Saxena et al. 1980).

Standardisation: A TLC method was developed for the standardization of *Ashokarishta* by quantitative estimation of major compounds, catachin, (+) catechole, (−) epicatechin (Parihar et al. 2010). RP–LC method for standardization of *Ashokarishta* has been developed (Govindarajan et al. 2008). A TLC method for standardization of different marketed brands of *Ashokarishta* has been developed (Kumar, Larokar and Jain 2013). β-sitosterol has been reported from the formulation (Gahlaut et al. 2013).

Pre-clinical research

Antioxidant: Antioxidant potential was assessed by free radical scavenging activity, superoxide radical scavenging activity and reducing power assay. Results revealed that the *Ashokarishta* formulated by Vaidyaratnam (Kerala) is a good source of antioxidant compounds followed by Baidyanath, where as samples from Rasashala-Pune and Sandu Brothers showed comparatively lower activities with aforesaid methods (Dushing and Shankar 2012).

Clinical research

Leucorrhoea **and *menorrhagia***: Several studies have been published proving efficacy of *Ashokarishta* in the treatment of leucorrhoea and menorrhagia (Geeta and Nalini 1996; Anonymous 2002).

Ashwagandhadi lehyam

Standardization: The physico-chemical studies showed total ash content as 6.45%, extractive values and some trace elements such as lead, mercury, cadmium and arsenic with 3.2, 0.05, 0.18 and 0.48 ppm, respectively. FTIR and HPTLC studies revealed the presence of functional groups of with anolides in *Ashwagandhadi lehyam*, resulting in its chemical standardization (Rasheed et al. 2013).

Preclinical pharmacology

Anti-epileptic: *Ashwagandhadi lehyam* exhibited less epileptic seizures in various phases when compared with that of standard phenytoin and was found to possess better anti-epileptic activity.

Ashwagandharishta

Preclinical pharmacology

Anticonvulsant: MES seizures were induced for rats and seizure severity was assessed by the duration of hind limb extensor phase. Pre-treatment with flax seed oil exhibited significant anticonvulsant activity by decreasing the duration of tonic extensor phase. Contrary to the expectations, pre-treatment with flax seed oil as an adjuvant to *Ashwagandharishta* failed to decrease the tonic extensor phase; however, it significantly decreased the flexion phase (P < 0.001) and duration of the convulsions (P < 0.05). Both the drugs exhibited an excellent anti-post-ictal depression effect and complete protection against mortality (Tanna et al. 2012).

Hepatoprotective: In the CCl4 induced hepato-toxicity model, *Ashwagandharishta*-2.31 ml/kg dose showed significant decrease in elevated hepatic level of AST (p < 0.001), ALT (p < 0.01) and ALP (p < 0.001). *Ashwagandharishta* administration revealed up-regulation in antioxidant genes such as CAT and GPx in liver with concomitant down-regulation in proinflammatory IL-6 gene ($p < 0.01$).

Asmarihara kasaya churna

Toxicology: The present study was designed for observing its effect on the liver function parameters on a rat's (albino rats -70–90 g) plasma after chronic administrations for 45 consecutive days. Various parameters that demonstrate liver functions were observed in both male and female rats (Bulbul and Choudhuri 2013).

Ashthamulika Taila

Clinical study

Filariasis: A clinical study explored the effect of external application of *Ashthamulika Taila* in 76 patients. Encouraging results were observed during and after 30 days of treatment. Out of 76 cases, 37 (48.68%) got good response, 24 (31.58%) got fair response, 8 (10.53%) got poor response and 7 (9.21%) cases did not show any response. 73.13% relief was found on over all acute clinical parameters (Prasad et al. 2011).

Apasmarari rasa

Preclinical pharmacology

Anti-epileptic: *Apasmarari rasa* was subjected to assess the LD 50 and Anti convulsant activity on male albino rats by means of Maximal Electro convulsing Shock Method. A supra maximal strength was 150 mA in rats for 0.2 seconds and stimulus was applied via ear clip electrodes. The animal dose of Phenytoin (7.2 mg/kg), Smriti sagar rasa (18 mg/kg) and *Apasmarari rasa* (5.4 mg/kg) was given orally to different groups. The animals were observed for a period of 180 minutes after being subjected to electro convulsions. Experimental study had shown some significant result when compared to other drugs (Saroch et al. 2012).

Avipattikar churna

Standardization: Energy dispersive X-ray spectroscopy has been used in setting quality control parameters of *Avipattikar churna* (Kumar and Nani 2012). The results obtained with the market formulations and the in-house formulations of *Avipattikar churna* were found to be comparable and variation was insignificant. Acid insoluble ash value for in-house formulation was found to be 0.356 ± 0.073 (Average value along with standard deviation), in case of marketed formulation, this was found to be 0.931 ± 0.160 and 1.197 ± 0.098 (Aswatha Ram et al. 2009).

Preclinical pharmacology

***Anti-Secretory* and *anti-Ulcerogenic*:** *Avipattikar churna* in both doses (500 mg/kg and 750 mg/kg), significantly decreased the volumes of the gastric contents, the ulcer score, the length of the ulcer, the gastric irritancy index and pH increased as compared to those in the control group. The effects of *Avipattikar churna* were comparable to that of ranitidine (Gyawali et al. 2013).

Balarishta

Preclinical pharmacology

***Anti-inflammatory*:** *Balarishta* was investigated for anti-inflammatory activity against cotton pellet induced granuloma in albino rats. There was significant reduction in cotton pellet weight and acid phosphatase, GPT and GOT activities by *Balarishta* (Alam et al. 1998).

***Antioxidant*:** Preliminary phytochemical screening showed the presence of carbohydrates, phenols, flavonoids, tannins, terpenoids, anthraquinone and fixed oil. *Balarishta* demonstrated antioxidant activity as evaluated by DPPH method (Rajalakshmy and Sindhu 2011).

Balachaturbhadra Avaleha

Pharmacognostical and Pharmaceutical Evaluation: Starch grain cells, calcium oxalate, tannin and fibres were the characteristic features observed in microscopic study of the formulation. The phyto-chemical evaluation of *Balachaturbhadra Avaleha* shows that the presence of carbohydrates, steroids, cardiac glycosides and flavanoids. The preliminary HPTLC study of the compound reveals the components are more sensitive to short UV 254 nm having 15 spots compared to long UV 366 nm with 09 spots (Joshi et al. 2014).

Balacaturbhadrika Churna

Toxicity: The study was carried out by administering *Balacaturbhadrika churna* orally once only in a dose up to 2000 mg/kg. For long-term toxicity, *Balacaturbhadrika churna* was administered in doses of 450 and 900 mg/kg orally for 45 consecutive days. Long-term toxicity results showed that, even at higher dose of 900 mg/kg, *Balacaturbhadrika churna* did not affect the parameters studied, to a significant extent. The doses employed for these toxicity studies were several times higher than normal clinical doses of *Balacaturbhadrika churna* (Nariya et al. 2011).

Bharangyadi Churna

Preclinical pharmacology

Antioxidant: Hydroalcoholic extract of *Bharangyadi Churna* resulted in dose depended increase in concentration maximum absorption of 0.677 ± 0.017 at 1000 µg/ml compared with standard Quercetin 0.856 ± 0.020. ABTS (+) assay shows maximum inhibition of 64.2 ± 0.86 with EC_{50} 675.31 ± 4.24. Superoxide free radical shows maximum scavenging activity of 62.45 ± 1.86 with EC_{50} 774.70 ± 5.45 (Kajaria et al. 2012).

Anti-histaminic, mast cell stabilizing and *bronchodilator*: Pre-treatment with *Bharangyadi* extract showed 80% & 86% protection from histamine induced bronchoconstriction in guinea pigs with 27.8% and 36.1% increase in preconvulsion time (equal to standard drug). Increasing concentration of *Bharangyadi* extract with maximum dose of histamine (1.6 µg) showed maximum inhibition at the dose of 50 mg (Kajaria et al. 2012).

Bhringaraja Taila

Standardisation: The initial investigations indicated 0.68% moisture, 0.46 acid value, 88.60 saponification value and 102.00 iodine value. Subsequent investigations after ten months indicate 0.90% moisture, 0.80 acid values, 190.50 saponification values and 04.00% iodine value (Tiwari et al. 1993).

Bilvadileha

Standardisation: Methanolic extract of *Bilvadileha* showed only piperine as a phytochemical constituent. The content of piperine in freshly prepared *Bilvadileha* sample was 0.72 ± 0.004 mg/g. Till three months of storage, no remarkable variation in the content of piperine in *Bilvadileha* was observed (Shailajan et al. 2013).

Clinical study

Irritable bowel syndrome: In present clinical trial 51 patients of Irritable bowel syndrome were registered out of which 46 patients completed the treatment. *Bilvadileha* was administered for the duration of 12 weeks. The therapy showed statistically significant improvement in all the clinical features of IBS as well as in the IBS severity scores (Tiwari et al. 2013).

Toxicity: In acute toxicity studies, no significant change in body weight, food intake and water intake of the animals was observed compared to the animals of the control group and also no mortality was recorded after the oral administration of *Bilvadileha* (Shailajan et al. 2013).

Bilvadi churna

Preclinical pharmacology

Wound Healing: In a study, it was found that wound healing time was significantly reduced when rats received *Bilvadi churna*. Also, there was no secondary infection in wounds in *Bilvadi churna* treated groups (Gosavi and Kale 2012).

Brahma Rasayana

Preclinical pharmacology

Antioxidant: Intraperitoneal administration of cyclophosphamide 25 mg/kg.b.wt. dose/mouse for 10 days was found to suppress the tissue and serum level of reduced glutathione, blood glutathione peroxidase and tissue levels of superoxide dismutase and catalase. Oral administration of *Brahma Rasayana* BR-50 mg/dose/mouse for 10 days and 30 days significantly enhanced the tissue levels of SOD, CAT, GST, GPX, serum and tissue GSH and significantly reduced the serum and tissue lipid peroxidation (Rekha et al. 2001).

The antioxidant properties of *Brahma Rasayana* (2 g/kg daily, orally) during cold stress was also studied. There was a significant ($P < 0.05$) decrease in antioxidant enzyme in the blood in cold stressed chicken. Serum and liver lipid peroxidation levels were significantly ($P < 0.05$) higher in cold stressed untreated chickens when compared to the treated and unstressed groups (Ramnath and Rekha 2009).

Anti-tumour: *Brahma Rasayana* treatment of MAT-LyLu cell inoculated Copenhagen rats resulted in a decrease of palpable tumor incidence, delay in tumor occurrence and lower mean tumor volumes. Also, a significant reduction in tumor weight and lung metastasis was observed in *Brahma Rasayana* treated animals in comparison to untreated controls (Thangapazham et al. 2006).

Nephroprotective: Administration of either *Brahma Rasayana* or *Chyavanaprash* (1 and 2 g/kg) maintained the antioxidant status in the kidney thereby preventing tissue damage as well as the release of marker enzymes. Cisplatin (Cis-dichlorodiammineplatinum) [II] induced variation of renal architecture was also prevented by *Brahma Rasayana* and *Chyavanaprash* administration (Menon and Nair 2013).

Nootropic: *Brahmi rasayana* (100 and 200 mg kg (-1) p.o.) was administered for eight successive days to both young and aged mice. Piracetam (200 mg kg (-1) i.p.) was used as a standard nootropic agent. *Brahmi rasayana* significantly improved learning and memory in young mice and reversed the amnesia induced by both scopolamine (0.4 mg kg (-1) i.p.) and natural aging (Joshi and Parle 2006).

Radioprotective: Oral administration of *Brahma Rasayana* (10 and 50 mg/dose/animal) for 15 days increased significantly total leukocyte count and percentage of polymorphonuclear cells in irradiated mice. Bone marrow cellularity and α-esterase positive cells also increased significantly in radiation-treated animals after BR administration. Oral administration of BR also enhanced in serum level of interferon-γ

(IFN-γ), interleukin-2 (IL-2), and granulocyte macrophage-colony stimulating factor (GM-CSF) in normal and irradiated mice (Rekha et al. 2000).

Genotoxicity: Nine months older mice were orally fed with rasayana for eight weeks. The treated groups showed no signs of dose-dependent toxicity at the dosage levels tested. The body weight loss/gain and feed consumption were unaffected at tested doses. Furthermore, sperm abnormalities and chromosomal aberrations were insignificant in the treatment group when compared to controls. However, there was a marginal increase in sperm count in the *Brahma Rasayana* treated animals (Guruprasad et al. 2010).

Clinical study

Anabolic effect: *Brahma Rasayana* was given to 20 patients in the dose of 10 gm twice daily with milk for 45 days and follow up done every 15 days. Evaluation was done by means of objective and subjective criteria. Results were significant in most of the objective and subjective criteria (Sharma, Mishra and Mankotia 2012).

Brahmi ghrita

Analytical profile: Organoleptic features of coarse powder made out of the crude drugs were within the standards prescribed. Acid value was 0.16075, saponification value 184.17, Refractive Index value 1.467 at room temperature, Iodine value 26.715, Specific gravity at room temperature was 0.9133 (Jyoti et al. 2012).

Acid values of Sample A, B, and C of *Brahmi Ghrita* were 4.26, 4.03, and 4.03; the Saponification values of Samples A, B, and C of *Brahmi Ghrita* were 227.2, 230.01, and 230.01, and the Iodine values of Samples A, B, and C were 34.75, 35.88, and 35.88, respectively, and the Acid value, Saponification value, and Iodine value of *Purana Ghrita* were 1.57, 199.15, and 31.04, respectively (Yadav et al. 2013).

Standardization: HPTLC method has been developed for quantitative determination of bacoside A in *Brahmi Ghrita* (Jyoti et al. 2012). Standardization of *Brahmi Ghrita* has been done with special reference to its pharmaceutical study (Yadav and Reddy 2012).

Preclinical pharmacology

Anti-amnesic: *Brahmi ghrita* (400 and 800 mg/kg, p.o.) and piracetam-treated rats significantly reversed the effect of scopolamine in modified elevated plus maze, passive avoidance, and active avoidance tests. But there were no significant differences observed in antiamnesic activity of *Brahmi ghrita* and standard drug (Yadav et al. 2012).

Anti-epileptic: *Brahmi ghrita* was tested for its oral effectiveness in controlling pentylenetetrazole-induced seizures in male albino rats and was compared with benzdiazepam. Thirty-day pretreatment with both Brahmighritham and benzdiazepam served to make the rats more insensitive to epileptogenic events (Shanmugasundaram et al. 1991).

Clinical study

Attention Deficit Hyperactivity Disorder: In the pilot exploratory study, *Brahmi ghrita* showed 66% decrease in total ADHD score. In the therapeutic confirmatory study, only 16% improvement was seen with *Brahmi ghrita*, which was similar to methylphenidate, standard treatment for ADHD that was used as a comparator in the present study. No side-effects were reported in both studies (Bhalerao et al. 2013).

Depression: Total 42 patients fulfilling the DSM-IV criteria for major depressive episode were registered for the study. Out of 42, 35 patients completed the treatment. In these, total 25 patients were managed with *Brahmi ghrita* and 17 with placebo capsule. In the former group, 20 patients completed the treatment, whereas in the later, 15 patients completed treatment. When effect on H.D.R.S. was analyzed, *Brahmi ghrita* showed moderate improvement in 50% patients and mild improvement in 35% patients. Its total effect was statistically highly significant ($p < 0.005$) as compared to placebo (Deole and Chandola 2008).

Toxicity: In acute toxicity study, Swiss strain albino mice were administered orally *Brahmi Ghrita* doses of 1, 2.5 and 5 g/kg and observed for behavioural changes and mortality, if any. In sub chronic toxicity study, Charles Foster albino rats were administered two doses of *Brahmi Ghrita*, i.e., 400 and 800 mg/kg, p.o. for 30 consecutive days.

There was no mortality or abnormal behaviour observed in acute toxicity study in mice at all the three dose levels. In sub-chronic toxicity study, *Brahmi Ghrita* did not produce any significant changes in body weight and daily food and water intake of rats when compared to control group rats (Yadav et al. 2014).

Brahmi vati

Standardization: The HPLC analysis showed significant increase in amount of Bacoside A3 and Piperine in the in-house sample of *Brahmi vati* when compared with all three different marketed samples of the same. Results showed variations in the amount of Bacoside A3 and Piperine in different samples which indicate non-uniformity in their quality which will lead to difference in their therapeutic effects (Mishra et al. 2013).

Candanasava

Fermentation studies: *Candanasava* was prepared and studied to examine whether these claims are tenable. Maximum alcohol production (9.8%) in 30 days was reached in the earthen pot. With the progress of time beyond 30 days, there was loss of yield, alcohol and sugar. There was rich growth of fungi in the pots. *Candanasava* stored in glass bottles did not show any change in any of the measured parameters. There was no increase of alcohol with the prolonged storage contrary to the claims (Alam et al. 1984).

Clinical study

***Pyuria*:** 40 patients with a urinary tract infection were included in the study; *candanasava* was administered in a dose of 40 ml/day in two divided doses after meals and mixed with an equal amount of water for a total duration of six weeks; significant symptomatic improvement occurred; though the urine did not become free from pathogen organisms, the pus cells, red blood cells and casts almost disappeared, as well as the albumen; several hypotheses regarding the origin of these effects are brought forward (Goel and Singh 1991).

Chaturjat Churna

In the present study two batches of marketed and laboratory formulations were compared and evaluated as per Indian Pharmacopeia and World Health Organization guidelines. The result revealed that all batches were in close proximity with that of standard values (Jhanwar et al. 2013).

Chandrakala Rasa

Clinical study: A clinical study has been done to evaluate the efficacy of *Chandrakala rasa* on *madhumeha* (Gautam et al. 2014).

Chandraprabha Vati

Standardisation: A colorimetric method has been developed to standardise *Chandraprabha vati* (Alam et al. 1982).

Characterization: The importance of purification steps was studied through the analysis of raw material and the intermediates using X-ray fluorescence spectroscopy. The scanning electron micrographs of *Chandraprabha vati* tablets show the presence of pores as well as the plate-like objects. The formation of pores may be attributed to the evaporation of volatile molecules during heating (Safiullah et al. 2012).

Elemental analysis: The elemental analysis showed the presence of iron, chlorine, potassium, calcium, sodium and aluminum at greater than 1 wt % each, while other elements like magnesium, sulphur and phosphorus are present at <1 wt % each (Safiullah et al. 2015).

Preclinical pharmacology

***Anti-diabetic*:** An experimental study reported anti-diabetic activity of *Chandraprabha vati* (Parimi et al. 1995).

Antioxidant: Methanolic extracts of *Chandraprabha Vati* and Maha *yogaraja Guggulu* were good scavengers of all the radicals but there was a difference in the activity of the two formulations in different models. *Chandraprabha Vati* was a good scavenger of superoxide radical and *Maha yogaraja Guggulu* was efficient in scavenging nitric oxide, while both inhibited lipid peroxidation efficiently (Bagul, Kanaki and Rajani 2005a).

Diuretic: A study investigated diuretic effect of *Chandraprabha Vati* in rat conscious hydrated model. The onset of the diuretic action was very rapid (within 1 h) and so was the peak diuresis (within 1 h) but the effect was short lived (2 h) as furosemide (Ratnasooriya et al. 2014).

Prostatic hyperplasia: *Chandraprabha vati* at dosage of 100, 200 and 400 mg/kg orally for 21 days showed significant reduction of prostatic hyperplasia as indicated by decreased prostate weight and volume and hispathological observations. It also significantly inhibited carrageenan induced hind paw oedema in dose dependant manner (Dumbre et al. 2012).

Chyavanprash

Standardisation: A HPTLC densitometric method has been designed to determine antioxidant constituents (Ladha et al. 2008). A high-performance liquid chromatography method for the separation and quantitative determination of catechin, quercetin-3-O-rutinoside, syringic acid and gallic acid within a 35 min analysis has been developed (Govindarajan et al. 2007).

Preclinical pharmacology

Anabolic: 20 albino rats have been studied and were divided into two groups of ten each. Each rat of group I was given *Chyavanprash* 0.33 gms daily through Ryle's tube while control group did not receive any medicine. There was definite increase in body weight and retentions of 24 hours urinary and fecal nitrogen in the group I (Ojha et al. 1973).

Anti-amnesic: In the present study, *Chyavanprash* (at the dose of 1 and 2% w/w of diet) administered daily for 15 successive days in mice with memory deficits. The administration of Chy for 15 consecutive days significantly protected the animals from developing memory impairment. There was a significant decrease in brain thiobarbituric acid reactive substances and increase in glutathione levels after administration of *Chyavanprash* (2% w/w) (Bansal and Parle 2011).

Anti-cataract: Aqueous extracts of *Chyavanprash* (3% and 10%) were applied at 3, 10 and 20 h after hydrocortisone administration in the developing chick embryo. The incidence of advanced opacity (stage 4 and 5) was 33% after treatment with 3% extract, and 16.5% after treatment with 10% extract, compared with 96.6% in the control group. The aqueous extracts (3% and 10%) significantly prevented depletion of glutathione (Velpandian et al. 1998).

Anti-inflammatory: Flavonoids from three different preparations (C1, C2, and C3) of *Chyavanprash* were administered to different groups of albino rats (100 mg/kg, i.p.) and their effect on inflammation by carrageenan induced paw edema was studied. The results were demonstrated at 0 h, 1 h, 2 h, 3 h and 4 h time intervals. The C1 preparation showed more anti-inflammatory activity compared to C2 while C3 showed the least anti-inflammatory activity (Chavan 2011).

Genoprotective: 25 individuals matched with regard to age, sex and social status constituted the control. 20 gms of *Chyavanprash* was fed to bidi smokers for two months, twice a day. Bidi smokers were compared with *Chyavanprash*-fed bidi smokers. In *Chyavanprash*-fed bidi smokers as compared with bidi smokers Mitotic index, chromosomal aberrations, sister chromatid exchanges and satellite associations showed a significant decrease ($P < 0.01$) (Yadav et al. 2003).

Nootropic: In the present study, *Chyavanprash* was administered orally in two concentrations (1 and 2% w/w of the diet) for 15 successive days to 17 different groups of young (8 groups) and aged (9 groups) mice. The administration of *Chyavanprash* (1 and 2% w/w) for 15 consecutive days significantly improved the memory of aged mice when compared to young mice (Bansal and Parle 2011).

Clinical study

Hyperglycemia **and** ***hyperlipidemia***: Ten normal healthy adult male volunteers (age 20–32 years) participated in the 16-week study. They were placed randomly in either the *Chyavanprash* group (n = 5) or vitamin C group (n = 5). Those in the former received 15 g/d of *Chyavanprash* while those in the latter received 500 mg/d vitamin C during the first 8 weeks of the study. An oral glucose tolerance test and lipoprotein profile in peripheral serum samples was determined at 0 weeks, 4 weeks, 8 weeks, 12 weeks and 16 weeks. *Chyavanprash* reduced postprandial glycemia in the oral glucose tolerance test and reduces blood cholesterol level to a significantly greater extent than vitamin C (Manjunatha et al. 2001).

Safety study: In 10 healthy adults, the blood samples were collected early morning in fasting state and again after four hours of their routine activity on the first day (control). On third day and seventh day, after collecting the fasting samples, they were either administered 100 g of fresh *amla* or boiled *amla* pulp (boiled for 3 hours) randomly. The samples were again collected after 4 hours.

Feeding of fresh amla pulp raised vitamin C level by 58% while boiled *amla* increased it by 48%. Parallel to this, fibrinolytic activity was also increased by 94% and 83% respectively. This clearly shows that only about 10% of vitamin C is destroyed during boiling of *amla* pulp (processing of *Chyavanprash*), while most of it is preserved (Bordia et al. 1981).

Chitrakadi Vati

Standardization: A simple, rapid, precise, accurate and reproducible High Performance Thin Layer Chromatography densitometric method was developed. The separation was performed on TLC aluminium plates precoated with silica gel 60 F254, good separation was achieved in the mobile phase of toluene:ethyl acetate:formic acid (7.5:2.5:0.5 v/v/v) and densitometry determination of piperine (0.39 ± 0.02) and plumbagin (0.70 ± 0.02) was carried out at single wavelength scanning at 280 nm (Khanvilkar et al. 2014).

Preclinical pharmacology

Antibacterial: The antibacterial activities of extracts of *Chitrakadi vati* in concentrations (5, 25, 50, 100, 250 μg/ml) were tested against the Gram negative – *Escherichia coli*, human pathogenic bacteria, occurring commonly in intestines. The results showed that the remarkable inhibition of the bacterial growth was shown by *Chitrakadi vati* against the tested organism. The antibacterial potential of the extracts were found to be dose dependent (Kamble and Dwivedi 2013).

Dashanga Agada

Physicochemical Analysis: Fluorescence analysis results indicated no fluorescent material in formulation. Microbial limit test showed there was no growth of organisms after 24 hrs of incubation as per Indian Pharmacopoeia. Thin Layer chromatographic analysis showed 10 and 11 picks at 254 nm and 366 nm respectively (Patil et al. 2013).

Dashanga Kwatha Ghanatablet

Standardization: The *Dashanga Kwatha Ghana* tablet was subjected to organoleptic analysis, phytochemical analysis, and qualitative analysis to detect the presence of various functional groups, and to high performance thin layer chromatography examination by optimizing the solvent systems. The investigation revealed the presence of tannins, mucilage, ascorbic acid, alkaloids, saponins, glycosides, flavonoids and carbohydrates (Baragi et al. 2011).

Dashanga Yoga

Anti-inflammatory: *Dashanga Yoga* significantly inhibited carrageenan-induced paw edema ($P < 0.01$) at both three and six hours; however, it failed to suppress formalin-induced paw edema. It decreased the formation of granulation tissue non-significantly in the chronic inflammatory model. In the tail flick model, *Dashanga Yoga* significantly increased tail flick latency at 120, 180, and 240 minutes in comparison to the control group (Ruknuddin et al. 2013).

Dasamularishta

Standardization: *Dasamularishta* was prepared and analysed for total alkaloidal contents in addition to the parameters prescribed in the Ayurveda pharmacopeia. The total alkaloids present in *Dasamularishta* were 0.1% (Alam et al. 1989).

Fermentation studies: *Citraka* (*Plumbago rosea* Linn.) was impoverished by 50% with respect to plumbagin as a result of sodhana-purification. The drug prepared in glass vessel showed higher amount of alcohol than the earthen pot product. During the process of fermentation two yeasts and one bacterium, *Micrococcus luteus* were observed in the media, the bacterium being a non-alcohol producing organism (Alam et al. 1984).

Dasamularishta was fermented in different volumes in earthen pots of identical size, shape and capacity, as well as in a stainless steel vessel and a porcelain jar. The drug filled up to 3/4(th) of the volume of the earthen pot had shown better results than the earthen pots containing various volumes of drug. The stainless steel container and porcelain jar also showed comparable results to the earthen pot fermented drug. Thin layer chromatography of different preparations showed five spots (Alam et al. 1988).

Quality assessment of marketed brands: The present investigation reveals that all the preparations contained acceptable levels of alcohol (less than 12% v/v). However, the preparations were found to contain unacceptable limits of microbial load although all showed the absence of *Escherichia coli, Salmonella* species, and *Staphylococcus aureus* (Kalaiselvan et al. 2010).

Dasamula Extract

Pre-clinical pharmacology

Anti-inflammatory: Albino Wistar rats of either sex were allocated to two groups of 24 animals each. Each group was further subdivided into four sub groups, receiving either saline as control or *Dashamula* low dose or *Dashamula* high dose or Diclofenac sodium. Statistical analysis was done by using Student's 't' test. Highly significant ($p < 0.001$) paw edema reduction in acute inflammation test and decrease in final weight of the cotton pellet in chronic inflammation test was seen in groups treated with *Dashamula* high dose and Diclofenac sodium and significant ($p < 0.05$) difference in Dashamula low dose in both the models as compared to the control (Singh et al. 2011).

Dasamula Kwatha

Standardization: *Dasamula Kwatha* was prepared and analysed for total alkaloidal contents in addition to the parameters prescribed in the Ayurveda pharmacopeia. The total alkaloids present in *Dasamula Kwatha* were 0.09% (Alam et al. 1989).

Pre-clinical pharmacology

Analgesic: The study indicates that *Dasamula Kwatha* extract effectively produced like aspirin analgesic effect. The aspirin-like effect is further supported by the significant anti-pyretic effect and mild anti-inflammatory effect in rats against carrageenin induced oedema (Gupta, Singh and Singh 1989).

Antinociceptive: Experiments were carried out on 'albino rat model' using tail flick method. *Dasamula Kwatha* in a dose of 1.0 g/100 g, found to highly significant analgesic. Tramadol showed no antinociception action when given in a dose 10 mg/kg i.p., but it showed excellent antinociception by increasing the tail flick latencies at affective dose, i.e., 25 mg/kg i.p. (Gehlot et al. 2008).

Dasamula Taila

Standardization: *Dasamula Taila* was prepared and analysed for total alkaloidal contents in addition to the parameters prescribed in the Ayurveda pharmacopeia. The total alkaloids present in *Dasamula Kwatha* were 0.16% (Alam et al. 1980; Alam et al. 1989).

Dhanvantara Gutika

Preclinical pharmacology

Anti-inflammatory: *Dhanvantara Gutika* was investigated for anti-inflammatory activity against cotton pellet induced granuloma in albino rats. There was significant reduction in cotton pellet weight and acid phosphatase, GPT and GOT activities by *Dhanvantara Gutika* (Alam et al. 1998).

Dhatri Lauha

Pharmacognostical and pharmaceutical evaluation: The pharmacognostical work reveals that presentation of Rhomboid crystal of *Yastimadhu* and Stone cell of *Amalaki*. Organoleptic features of coarse powder made out of the crude drugs were within the standard range as per mentioned in classic. The pH value of *Dhatri Lauha Vati* was 4.18, Water soluble extract was 20.2% w/w, Alcohol soluble extract was 14.9% w/w, Ash value was 48.17% w/w, Loss on drying was 3.12% w/w (Rupapara et al. 2013).

Pre-clinical pharmacology

CNS effects: A total of four experiments were carried out at different doses (100, 200 and 400 mg/kg, p.o.) of *Dhatri Lauha* in different animal model in an attempt to confirm the safety of the general patients. In Hole cross test highly significant ($p < 0.005$) increase in motor activity was observed but only at a dose of 200 mg/kg and only at min 30. Highly significant ($p < 0.005$) increase in ambulation behavior was observed in Hole board test only at a dose of 100 mg/kg after 240 min and a significant increase ($p < 0.01$) was observed at the same dose after 180 min. Climbing out test also did not produce any significant changes in activity (Nishat et al. 2012).

Clinical study

Hyperacidity: A clinical study reported beneficial effect of *Dhatri Lauha* in reducing symptoms of hyperacidity (Das and Shaw 1983).

Iron deficiency anaemia: A study conducted by Central Council for Research in Ayurveda and Siddha (CCRAS) reported efficacy of *Dhatri Lauha* in treatment of iron deficiency anaemia (Lavekar 2010). A Multi centric Open Observational study justified role of *Dhatri Lauha* in treatment of iron deficiency anaemia (Bhuyan and Srikanth 2010).

Total 26 patients were randomly divided in to two groups; Group A (n 12) Pandughni Vati 2 tablets of 250 mg tds and Group B (n 10) *Dhatri Lauha Vati* 1 tablet of 250 mg tds. Group A: The result observed in dyspnoea (60%) and palpitation (53.33%) were highly significant statistically (<0.001). Weakness (33.33%), fatigue (40%), anorexia (28.57%) and cramps (55.55%) were decreased significant statistically (<0.05) whereas in pallor (24%) it was not significant. In Group B, results observed were highly significant statistically (<0.001) in pallor (50%) and dyspnoea (56.25%). The results in fatigue (61.54%), palpitation (55.55%), anorexia (42.85%), cramps (49.49%), were significant (Rupapara and Donga 2012).

The 60 iron deficiency anemia pregnant women age group between 18 to 35 years divided into two group Group A (n = 30) and Group B (n = 30) were treated with *Dhatri lauha* and *Navayasa lauha* respectively. *Dhatri Lauha* provided significant result on Hb gm%, RBC, MCV, PCV serum iron percent transferrin saturation and TIBC where as insignificant changes were found in MCH and MCHC. *Navayasa Lauha* provided significant result on Hb gm% (Khot et al. 2013).

Drakshadi vati

Standardisation: A study reported studies on the standardisation of *Drakshadi vati* (Alam et al. 1985).

Drakshavaleha

Pre-clinical pharmacology

Cyclophosphamide induced weight loss: Loss of body weight due to *cyclophosphamide* administration could be minimized by feeding *Drakshavaleha* (16 g/kg) orally to pregnant mice from day "0" to day "18" of their gestation. Recovery observed in terms of body weight of the pups was statistically significant ($P < 0.001$)

in *Drakshavaleha* treated pups. *Drakshavaleha* also recovered the crown-rump length of the pups occurred due to cyclophosphamide administration (Kumar, Singh and Reddy 2013).

Drakshasava

Fermentation studies: A saccharomyces sp. causing fermentation in *Maducasava* was isolated from Maduca flowers (*Maduca longifolia*). The capability of the organism to cause fermentation in *Drakshasava* was carried out. The Saccharomyces sp. caused alcoholic fermentation and the alcoholic fermentation and the alcohol produced was almost equal in quantity to that prepared according to textual method. The change in the pH of fermenting medium did not help in increased production of alcohol (Muzaffer et al. 1983).

Draksarista

Fermentation studies: *Draksarista* was prepared on an experimental scale and the progress of fermentation was studied with reference to the conversion of sugars into alcohol and the attended variations in physico-chemical constants in the medium. It has been observed that a species of Saccharomyces isolated from Madhuka flowers was capable of causing fermentation in the *Draksarista* medium. However, its performance was less as compared to that of a Bacillus species isolated from Jaggery (Alam et al. 1983).

Pre-clinical pharmacology

***Diuretic*:** Oral administration of *Draksharishta*-T (prepared according to AFI), *Draksharishta*-M (dhataki flowers were replaced by yeast for inducing fermentation) and its marketed formulation at the dose of 2.0 ml/kg over a period of 5 h showed a significant increase in urine volume as compared to control group. Both types of *Draksharishta* showed significant increase in sodium, potassium and chloride level in urine sample as compared to control group. The maximum diuretic effect was produced by furosemide (Tiwari and Patel 2011).

Eladi Gutika

Pharmacognostic evaluation: Preliminary phytochemical evaluation revealed the presence of carbohydrates, alkaloids, flavonoids, essential oils, glycosides and tannins. A comparative evaluation of *Eladi Gutika* prepared in-house was carried out with three available marketed samples in terms of their respective piperine content (Shailajan et al. 2011).

Kinetic studies: A study has been performed for kinetic measurements of disintegration of *Eladi gutika* (Saxena et al. 1985).

Acute toxicity: Acute toxicity study of *Eladi Gutika* in Albino Swiss mice revealed that it is safe at the dose of 2 g/kg body weight in animal.

Gandarvahasthadi Kwath

Analytical study: While developing fingerprints of High-Performance Thin Layer chromatography study, four peak values at areas at % 8.05, 6.75, 9.32 and 5.89 were found in the HPTLC graph (Thundiparambil et al. 2012).

Gandhaka Rasayana

Chemical composition: *Gandhaka Rasayana* contains Fe (433.84), Cu (32.77), Ni (10.32), Zn (10.32), Ca (0.84%), Na (0.17%), As (12.14), triclinic cells, carbohydrates, steroids and tannin.

Pre-clinical pharmacology

Antifungal: For the study Broth Dilution method was adopted to study the Minimum Inhibitory Concentration and Minimum Contact Time of the *Gandhaka Rasayana*, following NCCLS guidelines. *Gandhaka rasayana* showed antifungal effect against Candida albicans with Minimum inhibitory concentration of 5.00 mg/ml of Brain Heart Infusion Broth and Minimum Contact Time above one hour (Prasanna Kumar et al. 2010).

Antimicrobial: *Gandhaka Rasayana* solutions in different concentrations showed a significant zone of inhibition against three strains of bacteria (22–30 mm) and four strains of fungi (16–25 mm) when compared to Fluconazole (22 mm) and control (Saokar et al. 2013).

Clinical study

Leucoderma: A study regarding efficacy of concomitant therapy of *Gandhaka Rasayana* with *Arogyavardhini vati* in the treatment of leucoderma has been published (Shetty et al. 2000).

Godanti Bhasma (Gypsum Calx)

Standardisation: Pharmaceutical standardisation of *Godanti bhasma* has been described (Vishwapal and Handa 1958; Nageswar and Dixit 1996).

Clinical study

Bronchial asthma: A combination of *Nardeeya Laxmi Vilas Rasa* and *Godanti Bhasam* has been found useful in cases of chronic bronchial asthma.

Non-specific lucorrhoea: Combination of *Nagkesara Churna* and *Godanti bhasma* has been successfully used in the treatment of non-specific leucorrhoea (Das and Hazra 2012).

Rheumatoid arthritis: Several papers regarding efficacy of *Godanti bhasma* in the management of Rheumatoid arthritis have been published (Kumar and Kumar 2003; Rao et al. 2006; Babu et al. 2009).

Gojihvadi Kwath

Stability study: The results of the study showed no marked changes in organoleptic as well as in physico chemical parameters indicating accelerated shelf life of granules over kwath preparation up to a period of six months (Neetu et al. 2011).

Gomutra Haritaki vati

Standardisation: The weight variation of In-house vati, marketed vati and marketed tablet were found to be 0.62 ± 0.01, 0.61 ± 0.16 and 0.53 ± 0.18 respectively. The hardness of In-house vati, Marketed vati and Marketed tablet were 10 ± 0.02 Kg/cm2, 17 ± 0.01 Kg/cm2 and 11 ± 0.01 Kg/cm2 respectively (Kadam et al. 2012).

Gokshuradi Guggulu

Pre-clinical pharmacology

Antimicrobial: Methanol & Ethyl Alcohol extracts of *Gokshuradi Guggulu* showed significant antimicrobial activity against all test organisms, Aqueous, Petroleum ether & Dichloromethane extracts of *Gokshuradi Guggulu* were found sensitive against *Escherichia Coli, Proteus Vulgaris, Staphyto-coccus Epidermidis* and were comparable to that of norfloxacin (Simpi et al. 2004).

Clinical study

Filariasis: 50 patients of chronic manifested filariasis were treated with *Kanchnar Guggulu* and *Goksuradi Guggulu*. Considering prognosis of the disease, the result seems encouraging. Since only a four week course of treatment 32% of patients had good response, 44% patients had fair response, 16% had poor response, 2% had no response and 6% were dropped out (Singh et al. 2008).

Gokshuradi Vati

Clinical study

Benign Prostatic Hyperplasia: In group A, *Gokshuradi Vati* 500 mg was given three times a day with luke-warm water after food; while in group B, *Dhanyaka-Gokshura Ghrita* as *Matra Basti* of 60 ml, once in a day, just after lunch and combined therapy of both formulations in group C was administered. Out of 32 patients, total 30 patients (10 in each group) were completed the treatment course of 21 days. In results, 54.09% improvement was seen in group C, 45.67% in group A and 47.99% in group B. The size of prostate gland was found reduced by a highly significant amount in group C (Bhalodia et al. 2012).

Gokshuradi Churna

Standardization: One marketed and one in-house formulation of *Gokshuradi churna* were used for the study. All the formulations were standardized on the basis of organoleptic characters, physical characteristics and physico-chemical properties. The set parameters were found to be sufficient to evaluate the *churna* and can be used as reference standards for the quality control/quality assurance purposes (Kumar et al. 2011).

Guduchi Ghana

Pre-clinical pharmacology

Anti-inflammatory: Two samples of *Guduchi Ghana* were evaluated for anti-inflammatory activity using carrageenan induced paw edema model in rats. Animals were divided in three groups, having six animals in each. Group A received the test drug, Group B received the market sample at a dose of 50 mg/kg orally, while Group C (control group) received tap water. Reduction in edema was observed in Group A and B at 3 h interval by 33.06% and 11.71% respectively. Group A showed significant effects ($P < 0.05$) in comparison to control group (Patgiri et al. 2014).

Guduchi Ghrita

Pre-clinical pharmacology

Adaptogenic: The adaptogenic effect of three samples of *Guduchi ghrita* obtained from three different sources was studied in albino rats and compared with expressed juice of stem of *Guduchi*. *Guduchi ghrita* produced better effect in comparison to the expressed juice. *Solapur Guduchi ghrita* was found to produce significant inhibition of stress hypothermia and gastric ulceration. In hematological parameters '*Guduchi*' juice produced better reversal of the stress-induced changes in comparison to the test '*ghrita*' preparations (Savrikar et al. 2010).

Antipyretic: Seven groups of six animals were used for the experiment. The yeast induced pyrexia method was standardized first by injecting 12.5% yeast suspension (s.c.) followed by recording the rectal temperature at regular intervals. *Guduchi ghrita* samples significantly attenuated the rise in temperature after three hours of yeast injection.

Guduchyadi Kwatha

Physico-chemical evaluation: Comparative organoleptic screening of fresh *Guduchyadi Kwatha* (GKF) and *Guduchyadi Kwatha* for instant use (GKI) showed no major differences in color and smell. In GKF, pH value, total solid content, specific gravity, and surface tension were found as 6.0, 95.14%, 1.009 w/w and 27.19 dynes/cm respectively. In finished product (GKI), pH value, loss on drying, Ash value, water soluble and alcohol soluble extractives were obtained as 6.16, 1.93% w/w, 7.90% w/w, 63.69% w/w and 37.29% w/w respectively (Prajapati, Sharma and Prajapati 2013).

Guduchi Satva (Starchy substance from *Tinospora cordifolia*)

Seasonal variations in physicochemical profile: 18 batches of *Guduchi Satva* were prepared in six different seasons (3 batches in each season) and findings were systematically recorded. Maximum yield of *Guduchi Satva* was obtained in January-February while the minimum was in May-June. Variation in taste and color was found in *Guduchi Satva* prepared in rainy season. All functional groups were found to be same in each season. Total alkaloidal contents were found a bit higher in rainy and spring seasons (Sharma et al. 2013).

Quality control evaluation: Fifteen batches of *Guduchi Satva* were prepared and findings were systematically recorded. The average percentage of dried *Guduchi Satva* obtained was 3.8%. Alkaloids, carbohydrates and starch were found present in *Guduchi Satva*. Number of peaks obtained in HPTLC also corresponds to this finding. Percentage of total alkaloid content was 0.31%. No heavy metal and microbial load were detected in the sample (Sharma et al. 2013a).

Pre-clinical pharmacology

Antioxidant: The present study was concentrated on the *in vitro* antioxidant methods where *Guduchi Satva* and hydro-alcoholic extract of *Curculigo orchioides* screened for DPPH free radical scavenging activity, total reducing power assay and hydrogen peroxide scavenging activity assay. The results revealed potent scavenging activity when compared with standard (Onkar et al. 2012).

Hypoglycemic and *anti-hyperglycemic*: *Guduchi Satva* was suspended in distilled water and administered to animals at the dose of 130 mg/kg that showed the marginal reduction in blood sugar level at all the time intervals in normoglycemic mice. In anti-hyperglycemic activity, administration of *Guduchi Satva* prior to glucose over load failed to attenuate blood sugar level at all-time interval in comparison to glucose control group (Sharma et al. 2013).

Hajrul Yahud Bhasma (Fossil stone calyx)

Pre-clinical pharmacology

Lithontriptic: Male Wistar albino rats (250–300 g) were selected and divided into five groups. Except group I (normal) and group-II (Sham operated), in group-III (model control), group-IV (standard) and group-V (test) urolithiasis was induced by Calcium oxalate seed deposition in the bladder surgically (3 mm diameter). Cystone (750 mg/kg, p.o.) and *Hajrul Yahud Bhasma* (100 mg/kg and 200 mg/kg, p.o.) were administered to group-IV and group-V for 14 days after surgery respectively. X-rays were taken before and after the treatment.

Increase in percent matrix growth, stone forming promoters and decrease in physical parameters and stone forming inhibitors induced by surgical implantation of Calcium oxalate seed was significantly reversed by treatment with *Hajrul Yahud Bhasma* (Rao et al. 2011).

Haridradi Ghrita

Preclinical pharmacology

Hepatoprotective: *Haridradi ghrita* (50, 100, 200 and 300 mg/kg), significantly lowered marker enzymes (SGPT, SGOT, ALP) and bilirubin in serum and liver peroxide, superoxide dismutase and catalase in liver homogenate following CCl_4 (0.7 ml/kg, ip) toxicity. The protective effect was further supported by reversal of CCl_4 induced histological changes (Satturwar et al. 2003).

Immunomodulatory: *Haridradi Ghrita* was administered orally at doses of 50, 100, 200 and 300 mg/kg/day to healthy rats showed a significant increase in neutrophil adhesion and delayed type hypersensitivity response whereas the humoral response to sheep RBCs was unaffected. With a dose of 200 and 300 mg/kg/day the DTH response (mean+SD % increase in paw volume) was 10.52 ± 3.12 and 14.50 ± 2.38 respectively, in comparison to the corresponding value of 6.01 ± 1.85 for the untreated control group (Fulzele et al. 2003).

Haridra Khanda

Standardization: The mean recovery values of 99.63 (curcumin), 98.65 (demethoxycurcumin) and 98.97% (bisdemethoxycurcumin) indicated the excellent accuracy of the method. It is shown that High-performance thin-layer chromatographic can be applied successfully for the marker evaluation of the formulation containing *C. longa* (Rout et al. 2008). A RP-HPLC method has been developed for determination of curcumin in *Haridra Khanda* (Chittora et al. 2010).

Clinical study

Allergic rhinitis: Total 32 patients were registered and randomly divided into two groups. In group A *Haridra khanda* and in group B *Pippalyadi taila Nasya* along with *Haridra khanda* were given for two months. The effect of therapy in both groups was assessed by a specially prepared proforma. In oral group and combined group, maximum number of patients, i.e., 45.45% and 53.33% respectively showed marked improvement (Bhakti et al. 2009).

Urticaria: The results of clinical study conducted with *Haridra Khanda* in 20 patients diagnosed with were assessed and found statistically significant (P > 0.01) (Swamy et al. 1999).

Hartala Bhasma

Preclinical pharmacology

Antimicrobial: *Hartala bhasma* has an effective antimicrobial activity against *Streptococcus pneumoniae, Klebsiella pneumoniae, Pseudomonas aeruginosa* and *Staphylococcus aureus* in both Diffusion and Dilution methods (Garg et al. 2013).

Hartala godanti bhasma

Preclinical pharmacology

Antimicrobial: *Hartala godanti bhasma* has an effective antimicrobial activity against *Streptococcus pneumoniae, Klebsiella pneumoniae, Pseudomonas aeruginosa* and *Staphylococcus aureus* in both Diffusion and Dilution methods (Garg et al. 2013).

Hinguadi taila

Standardization: *Hinguadi taila* had a specific gravity of 0.908, volatile oil content of 2.251%, saponification number 178.6, iodine number 105.4 and 0.2% allylisothiocyanate. On thin layer chromatography, Hinguadi taila resolved into six sulphuric acid positive spots, out of which three were derived from mustard oil and one from ginger (Alam et al. 1988).

Hinguleshwar Rasa

Clinical study

Influenza: In a comparative clinical study involving 60 patients, 31 were administered *Tribhuvan Kirti Rasa,* 28 were administered *Hinguleshwar Rasa* and to 1 both the drugs. *Hinguleshwar Rasa* and *Tribhuvan Kirti Rasa* were effective in Influenza but *Hinguleshwar Rasa* was found more effective (Pandey 1968).

Hingusauvarchaladi Ghrita

Preclinical pharmacology

Anticonvulsant: In the present study, *Hingusauvarchaladi ghrita* and *Saptavartita Hingusauvarchaladi ghrita* were studied for anticonvulsant activity experimentally on albino mice, by the chemoshock method. The observations recorded have been analyzed by one-way ANOVA followed by Scheffe's test, statistically. *Saptavartita Hingusauvarchaladi ghrita* has shown better anticonvulsant activity in comparison to *Hingusauvarchaladi ghrita* (Roshy and Ilanchezhian 2010).

Hiraka Bhasma

Composition

(A) Major constituents are carbon, oxygen, iron and silica.
(B) Minor constituents are potassium, calcium, magnesium and aluminium and
(C) Trace constituents are sodium, phosphorus and sulphur (Jawale et al. 2007).

Indrayanadi Yog

Clinical study

Diabetes mellitus: 43 patients were randomly selected for the study and administered *Indrayanadi Yog* for sixty days. After completion of treatment, the results were assessed by using following criteria: FBS and PPBS level, HbA1c value. After the treatment, it is observed that there is 27.76% reduction in Fasting and post-pandial blood sugar level, reduction in HbA1c (GHB) level clinical symptoms (Behera et al. 2013).

Jatyadi Ghrita

Standardization: Quantitative estimation was done by reported RP-HPLC methods using markers glycyrrhetinic acid, ursolic acid, karanjin, curcumin, berberine and kutkin which are major constituents of formulation. A standard laboratory reference sample of *Jatyadi Ghrita* and four marketed samples were evaluated as per methods (Vite et al. 2013).

Pre-clinical pharmacology

Anti-inflammatory: An experimental study reported efficacy of *Jjatyadi ghrita* in suppressing inflammation (Fulzele et al. 2002).

Jatyadi Taila

Phytochemical evaluation: Flavonoids, essential oils, tannins, glycosides, steroids and alkaloids. HPTLC confirms the presence of karanjin, lupeol and β-sitosterol.

Pre-clinical pharmacology

Wound-healing: Topical application of *Jatyadi Taila* on excision wounds caused significantly faster reduction in wound area as compared to the application of Neosporin and untreated control wounds. Animals treated with *Jatyadi Taila* showed significant increase in protein, hydroxyproline and hexosamine content in the granulation tissue when compared with the untreated controls. Wound healing potential of *Jatyadi Taila* was found to be dose dependant (Shailajan et al. 2011).

Jeevantyadi Ghrita

Clinical study

Myopia: A total of 54 patients (108 eyes) having myopia $\geq$-6 D were registered for the study and divided into two groups. Group A, *Akshi-Tarpana* with *Jeevantyadi Ghrita*, and Group B, *Akshi-Tarpana* with plain *Go Ghrita*, by stratified sampling. The procedure was done in 5 sittings on 5 days each with an equal interval of 5 days between each sitting.

A total of 22 patients in Group A and 18 in Group B completed the treatment. Obtained data were statistically analyzed using a t-test and the study reveals that objectively, 09.30% and 05.55% eyes were cured, 16.28% and 02.78% markedly improved, and 34.88% and 11.11% moderately improved in Group A and B, respectively (Manjusha et al. 2011).

Jwarhar mahakashay

Chemical investigation: The aqueous extract of *Jwarhar mahakashay* containing *Hemidesmus indicus* R. Br., *Rubia cordifolia* L., *Cissampelos pareira* L.; fruits of *Terminalia chebula* Retz., *Emblica officinalis* Gaertn., *Terminalia bellirica* Roxb., *Vitis vinifera* L., *Grewia asiatica* L., *Salvadora persica* L. and granules of *Saccharum officinarum* L. is used as a antipyretic. Gas chromatography-mass spectrometry led to the identification of 2-(1-oxopropyl)-benzoic acid as principal constituent (Gupta, Shaw and Mukherjee 2010).

Kanchnar Guggulu

Filariasis: 50 patients of chronic manifested filariasis were treated with *Kanchnar Guggulu* and *Goksuradi Guggulu*. Considering prognosis of the disease the result seems encouraging. Since only a four week course of treatment 32% of patients had good response, 44% patients had fair response, 16% had poor response, 2% had no response and 6% were had dropped out (Singh et al. 2008).

Psoriasis: 36 patients diagnosed with psoriasis were taken up for study. Following were prescribed

1. *Kaishore guggulu* – 1 gm mixed twice daily with Laghu Mangisthadi kwath 25 gm (Decoction form).

2. *Kanchnar guggulu* – 1 gm.
3. *Arogyavardhini vati* – 250 mg.

For external use *Kajali Kodaya malhar* was also used. 50% of patients showed encouraging improvements (Jha and Pandey 1985).

Kajjali

Physicochemical characteristics: A study evaluated the physicochemical characteristics of three samples of *Kajjali* prepared using three different samples of mercury (one extracted from Cinnabar and two other samples available in the market). Physicochemical analysis of all three samples of Kajjali showed similar readings. In X-Ray Diffraction analysis, all three samples had identical peaks, indicating similar chemical composition and Scanning Electron Microscope analysis indicates similar particle sizes (Santhosh et al. 2012).

Sub-acute toxicity: Different doses of *Kajjali* powder suspension were given to the animals for a period of 60 days. The changes in hepatic markers, lipid profile, renal markers and other behaviors were noted. From the results, it was concluded that *Kajjali* powder suspension has no acute and subacute toxicity (Louis et al. 2014).

Kalpita Yoga

Clinical study

Ischaemic Heart Disease: In a clinical study, it was observed that there was significant improvement in clinical features of *Ischaemic Heart Disease* after the therapy with *Kalpita Yoga*. The level of S. cholesterol, LDL, VLDL, S. triglycerides, Total lipids decreased and the level of HDL increased significantly after therapy (Sharma et al. 2002).

Kamdudha rasa

Clinical study

Acute promyelocytic leukemia: A 47 year old diabetic male patient was diagnosed and treated for high risk AML-M3 at Tata Memorial Hospital (BJ 17572), Mumbai in September 1995. The patient was given two cycles of chemotherapy with Idarubicine and Etoposide, after which he achieved remission. His disease again relapsed in December 1997. The patient was treated with a proprietary Ayurvedic medicine *Navajeevan, Kamadudha Rasa* and *Keharuba Pisti* for one year. For the subsequent 5 years the patient received three months of intermittent Ayurvedic treatment every year (Prakash et al. 2010).

Duodenal ulcer: A clinical trial comprising of *sutasekhara rasa, dhatri lauha* and *kamdudha rasa* was reported to be efficacious in the treatment of duodenal ulcer (Dash et al. 1989).

Kanakasava

Standardization: Method of preparation was optimized for pH and CO_2 release as check parameters for completion of fermentation and prepared formulation was standardized for preliminary physico-chemical parameters as per Ayurvedic Pharmacopoeia of India. The results of standardization, i.e., pH (3.86 ± 0.029), specific gravity (1.046 ± 0.009), viscosity (1.52 CS ± 0.006), phenolic content (0.079% w/v $\pm 0.012\%$), alcohol content (7.18% v/v ± 0.577), total solid content (14.64% w/v ± 0.348) and toxicological determinants complies with the official limits (Arora and Ansari 2013).

Pre-clinical pharmacology

Antimicrobial: The antimicrobial activity of *Kanakasava* was investigated using the agar well method. *Kanakasava* showed 17.5 mm zone of inhibition against Proteus sp., Enterococcus sp. (Chakrbortty and Ahmed 2013).

Kankayana vati

Clinical study

Hemorrhoids: A study reported efficacy of concomitant therapy of *Kankayana vati*, *Triphala churna* and *Kasishadi taila* in the management of anorectal piles (Singh et al. 2005).

Kaphaketu Rasa

Clinical study

Migraine: Patients diagnosed with migraine were divided in to three groups G1, G2, and G3 and prescribed *Anu taila nasya*, *Kaphaketu Rasa vati* orally and both to these groups respectively. The trial was conducted for a period of one month. The results were encouraging and maximum relief was observed in half sided headache, tinnitus and intolerance to light (Madaan and Sharma 2005).

Karpurasava

Standardisation: *Karpurasava* is prepared by distilling the alcohol. The quantity of alcohol in the distillate was 9.87% by the textual method while it was 5.77% by glass distillation apparatus. The alcohol content in *Karpurasava* was 8.7% (Alam et al. 1994).

Karpuradi Taila

Standardisation: The physicochemical standards and the Thin Layer chromatographic pattern can be used as a finger print standard for *Kapruradi Taila* (Hepsibah et al. 1998).

Karpura Shilajit Bhasma

Elemental composition: Calcium was found to be the major element (24.6044% w/w) by Inductively Coupled Plasma Atomic Emission Spectrometry.

Pre-clinical pharmacology

Diuretic: *Karpura shilajit bhasma* produced mild reduction in motor coordination and mild sedation during the first 2 h after administration at the doses of 500–5000 mg/kg p.o. in mice. The diuretic activity was found to be significant ($P < 0.01$) at minimum dose of 200 mg/kg p.o. in rats and dose dependant up to 1000 mg/kg p.o.

Kasishadi taila

Hemorrhoids: A study reported efficacy of concomitant therapy of *Kankayana vati*, *Triphala churna* and *Kasishadi taila* in the management of anorectal piles (Singh et al. 2005).

Kshatantak Malam

Pre-clinical pharmacology

Wound-healing: An 8-mm-diameter full-thickness punch was produced in Wistar rats. *Kshatantak Malam* was applied topically and compared with framycetin ointment and povidone iodine ointment. Significant results ($P < .05$) were observed with *KshatantakMalam* in the punch wound model on the basis of various physical, biochemical, and histopathological parameters. The drug was found to be safe in acute and chronic toxicity models in animals (Gangopadhyay et al. 2014).

Kayyonyadi Churna

Clinical study

Anemia: An open end clinical trial was taken up for the study on 30 patients and 1–2 gms of the churna is given twice daily along with *Takra* for 30 days. It was observed from the study that the drug showed marked reduction in the clinical symptoms (p value < 0.0001) and improvement in the Hb% (p value < 0.0001) (Prasanna Kumari 2013).

Keharuba Pisti

Clinical study

Acute promyelocytic leukemia: A 47 year old diabetic male patient was diagnosed and treated for high risk AML-M3 at Tata Memorial Hospital (BJ 17572), Mumbai in September 1995. The patient was given two cycles of chemotherapy with Idarubicine and Etoposide, after which he achieved remission. His disease again relapsed in

December 1997. The patient was treated with a proprietary Ayurvedic medicine *Navajeevan, Kamadudha Rasa* and *Keharuba Pisti* for one year. For the subsequent five years, the patient received three months of intermittent Ayurvedic treatment every year (Prakash et al. 2010).

Khadirarista

Pre-clinical pharmacology

Chronic administration: *Khadirarista* was studied for its effect on different hematologic parameters after chronic administrations for 45 days to male Sprague-Dawley rats. There was an (33.33%) increase in the percentage of monocyte count (3.55%) decrease in the hemoglobin, statistically significant (p = 0.028) increase in the number of platelet count and statistically very highly significant (p = 0.0) increase in the platelet volume distribution width of the male rat [3.31% increase] (Chakraborty et al. 2012).

Toxicology: *Khadirarista* was administered orally to the male albino rat. After 46 days of treatment period, animals were fasted for 18 hours after the last administration. Triglyceride and Creatinine levels were significantly (p = 0.001) decreased in the *Khadirarista* group than the corresponding control group. Bilirubin, Urea, sGPT, sGOT, ALT contents were also significantly (p = 0.001) decreased in KDR group.

Alcohol content: Estimated alcohol (as self generated alcohol) was found to be 8.1% in *khadirarishta* (Vora et al. 2012).

Koshashma (Asbestos calyx)

Clinical study

***Epilepsy*:** A clinical study reported efficacy of *Tantupasana*, a compound containing the bhasman of asbestos and a number of herbs in the treatment of epilepsy (Bapat 1982).

Tantu pashan on acute administration in maximal electrical shock induced seizures; only at the highest dose (160 mg/kg) significantly reduced the duration of hind limb tonic extension, while in Pentylenetetrazole induced seizures induced seizures, it did not alter the duration of seizures to any significant level when compared with control group.

On chronic administration, the test drug significantly reduced the duration of tonic hind limb extension and also the clonus phase in maximal electrical shock induced seizures. But, in Pentylenetetrazole induced seizures, not protected the animals from death (Sudhakar et al. 2010).

Kshiramandura

Analysis of Kshiramandura prepared according to traditional Ayurvedic method showed 68.35% Ferric oxide, 0.66% $MgCO_3$ and 1.32% $CaCo_3$ (Jadar and Jagadeesh 2010).

Kumaryasava

Standardisation: Various extracts of *Kumaryasava* have been prepared and evaluated. Chloroform extract indicated presence of three well-resolved fluorescent components. Spectral data of these three fractions (III–V) have been reported as a valuable analytical tool for routine standardization of *Kumaryasava*. Fraction V indicated presence of anthraquinones, which is reported as the main constituent of aloe, namely aloin (Elamthuruthy et al. 2005).

Pre-clinical pharmacology

Antioxidant: The aim of the present investigation is to find the effect of its preparation method on antioxidant activity of *Kumaryasava*. All the tested formulations showed marked *in vitro* antioxidant activity in which WFKA (*W. fruticosa* flowers based *Kumaryasava*) showed prominent activity. Obtained IC50 values of WFKA in DPPH scavenging assay, hydrogen peroxide scavenging assay and total reducing power assay were 481.78, 50.13 and 49.60, respectively (Manmode et al. 2012).

Hepatoprotective: Oral administration of Liv.52 and *Kumaryasava* to carbon tetrachloride (CCl4) treated rats improved growth. *Kumaryasava* was more effective in reducing the liver weight increase due to hepatotoxicity of CCl4. Hepatic arginase, cathepsin-B, acid phosphatase, ribonuclease activity, which were decreased on CCl4 treatment was stimulated by both Liv.52 and *Kumaryasava* (Kataria and Singh 1997).

Kushmandadi Ghrita

Clinical study

Generalized Anxiety Disorder: In survey study of 60 patients of Generalized Anxiety Disorder, significant symptomatic disturbance in mental health was observed. The patients divided in two groups were treated with *Kushmandadi Ghrita* and placebo control for four weeks. *Kushmandadi Ghrita* provided significant relief on Hamilton Anxiety Rating Scale, Brief Psychiatry Rating Scale and *Manasa bhava Pariksha* (Ahir et al. 2011).

Kutajarishta

Standardisation: Initial work on standardisation of *Kutajarishta* was carried out in 1965 (Vasudevan and Khorana 1965). TLC was developed as a tool for standardisation of *Kutajarishta* (Hiremath Shobha et al. 1993). A new spectrophotometric method has been developed for the estimation of total alkaloids in *Kutajarishta* (Niranjan et al. 2002).

Pre-clinical pharmacology

Antidiarrhoeal: The percentage inhibition of diarrhoea in castor oil treated animals were 48.54%, 71.27%, 86.36% and 92.45% for *Kutajarishta* 2.5, 5, 10 ml/kg dose (p.o.) and Loperamide (standard) 2 mg/kg dose (p.o.) respectively. *Kutajarishta* significantly reduced intraluminal fluid accumulation (Shamkuwar and Shahi 2012).

Amoebiasis: *Kutajarishta* and *kutajasava* (0.5 ml orally daily), administered to guinea pigs inoculated with *Entamoeba histolytica*, did not show any significant action on the caecal pathology; these findings are not in agreement with the symptomatic relief in patients with clinical amoebiasis; a notable observation was the significant lowering of the serotonin level in the blood of the guinea pigs (Dixit et al. 1988).

Laghugangadhar Churna

Standardisation: Four samples of *Laghugangadhar Churna* were procured from different Ayurvedic pharmacies and were investigated by microscopy, physico-chemical parameters and high performance thin layer chromatography. The microscopic analysis of samples revealed starch grains in three ingredients viz. *Cyperus rotundus*, *Symplocos racemosa* and *Woodfordia fruticosa*. HPTLC analysis of all four samples revealed the presence of all the ingredients in compound formulations as determined by simultaneous application of authentic ingredients (Rathi et al. 2010).

Laghupanchamula

Preclinical pharmacology

Anti-inflammatory: LPGE and LPEE (1.0 g/kg) at 3 h after their administration showed inhibition of formalin-induced paw edema by 46.2% and 44.3% ($P < 0.001$) and carrageenan-induced paw edema by 53.9% and 60.4% ($P < 0.001$), respectively. After 7 days of treatment, both LPGE and LPEE showed 26.3% ($P < 0.01$) and 32.5% ($P < 0.05$) inhibition, respectively, against formalin-induced paw edema, and reduced weight of turpentine-induced granuloma pouch by 42.8% and 36.1% ($P < 0.001$), and volume of exudates by 31.2% and 36.2% ($P < 0.001$), respectively (Ghildiyal et al. 2013).

Analgesic and hypnotic: Both the extracts (LPGE and LPEE) did not produce any acute toxicity in mice at single oral dose of 2.0 g/kg. Both LPGE and LPEE (250, 500, and 1000 mg/kg) showed dose-dependent elevation in pain threshold and peak analgesic effect at 60 min as evidenced by increased latency period in tail-flick method by 25.1–62.4% and 38.2–79.0% respectively. LPGE and LPEE (500 mg/kg) increased reaction time in hot-plate test at peak 60 min analgesic effect by 63.2 and 85.8% and reduction in the number of acetic acid-induced writhes by 55.9 and 65.8% respectively. Both potentiated pentobarbitone-induced hypnosis as indicated by increased duration of sleep in treated rats (Ghildiyal et al. 2014).

Lakshadi Guggulu

Clinical study

***Osteoarthritis*:** Ten patients having osteoarthritis of the knee were given *Lakshadi Guggulu, Kalka patra-bandhan* (Bandage of medicinal paste) and knee joint traction. The duration of treatment was one month with following follow-up every week. At the end of four weeks, statistically significant results were found in the criteria of assessment, specifically in severity of pain, deep grading of tenderness, walking distance and movement of knee joint (degree of flexion). Maximum response was observed in the deep grading of tenderness (Sharma, Sharma and Kushwah 2007).

For the present work, 30 clinically diagnosed patients were selected and randomly divided into three groups. Group A treated with *Laksha Guggulu* orally, Group B treated with snehana & swedana traction, Group C treated with *Laksha Guggulu, Snehana* (Oleation), *Swedana* (Fomentation) & Knee Joint Traction. Significant results were obtained on pain in joint movement, restriction in joint movement, joint stiffness, and local crepitation nearly in all the groups with best result in combined group or group C (Rajoria et al. 2010).

In a study, Group A treated with *Lakshadi guggulu* 500 mg twice orally, Group B treated with *Panchatikta Ksheer Vasti* (420 ml), Group C treated with *Lakshadi Guggulu* and *Panchatikta Ksheer Vasti*. The total duration of this work was 30 days with one month of follow up period. Significant result were obtained on pain in joint movement, restriction in joint movement, joint stiffness and local crepitation in nearly all the groups with the best result in combined group or group C (Singh and Rajoria 2014).

Lasunadi Guggulu

Clinical study

***Anti-anginal*:** During the clinical trial it was observed that there was significant improvement in clinical manifestations of stable angina after the therapy with *Lasunadi guggulu*. The level of S. Cholesterol, L.D.L., VLDL and serum triglycerides decreased and the level of HDL increased considerably after the therapy (Sharma et al. 2012).

Lasunadi vati

Standardisation: A study reported studies on the standardisation of *Lasunadi vati* (Sahoo and Swain 2011).

Lauha Bhasma (Iron calyx)

Standardization and bioavailability: Qualitative analysis indicates the presence of iron both in the ferric and the ferrous forms, a simple spectrophotometric method has been used for simultaneous determination of ferric ferrous and the total iron content in a single aliquot (Verma and Prasad 1995).

Randomly selected anaemic rabbits were divided into different groups and three variants of *Lauha bhasma* and ferrous sulphate sample were administered to each group. The effect of each formulation was monitored by measuring the haemoglobin content spectrophotometrically (cyanomethaemoglobin method). Increase in the haemoglobin content was found to be significant in case of the ayurvedic formulations as compared to the ferrous sulphate sample (Verma and Prasa 1995).

Chemical evaluation: Samples of marketed *Lauha bhasma* from different manufactures were evaluated chemically. Apart from the 81–85% iron content, the 15–19% other constituents were determined therein. Ferrous ferric and total iron in a single aliquot were determined spectrophotometrically, Qualitative and chromatographic analysis indicate the presence of sodium, potassium, calcium copper and cobalt in the samples, silicious matter and traces of ascorbic acid were present while tannin was absent in *Lauha bhasma* (Keshri et al. 1996).

Chemical characterization of *Lauha bhasma* by X-ray diffraction and vibrating sample magnetometry has been described (Singh et al. 2010).

Pre-clinical pharmacology

Haematinic and ***cytoprotective***: In Charles Foster strain rats of either sex, anemia was induced by administering mercuric chloride (9 mg/kg). *Lauha bhasma* and *Mandura bhasma* (11 mg/kg) were evaluated for their haematinic activity. Results suggest that *Lauha bhasma* and *Mandura bhasma* possess significant (P < 0.05) haematinic and cytoprotective activity (Sarkar et al. 2007).

Toxicity: *Lauhasava* treated mice of both sexes gained less weight than their control counterparts. The percentage of lung to the body weight was significantly increased in both sexes of rats. Liver weight in LSV treated rats of both sexes were observed to be increased. The percentage of kidney weight was increased in both sexes of rats, the result being significant in the case of female rats. Significant increase in the weight of rats' ovaries was observed (Ullah et al. 2008).

Lauha Parparti (Iron Flake)

Manufacturing: Purified mercury and sulphur, each in a quantity of 10 gm, were triturated well in a *khalvayantra* (mortar and pestle) to form a *kajjali*. 10 gm of *Kantalauha Bhasma* was added and trirurated for one hour continuously to form a homogeneous mixture; the mixture so obtained was then taken in a ghee-smeared iron ladle and heated on moderate fire till it melted completely; the melted material was poured on a ghee-smeared banana leaf, which was kept on a platform of cow dung and was immediately compressed by another ghee-smeared banana leaf for one minute; after cooling down the thin flake was obtained (Kotabagi et al. 2010).

Pre-clinical pharmacology

***Hepatoprotective*:** *Lauha parpati* was prepared by using four different methods and hepatoprotective activity was compared. The results showed that *Dvigunajarita*

Lauha parpati prepared by using *Kantaloha bhasma* shows significantly higher hepatoprotective activity compared to other *Lauha parpati* (Kotabagi et al. 2011).

Lavangadi Vati

Standardization: High Performance Thin Layer Chromatography method was developed for simultaneous quantitation of eugenol, piperine and ß-sitosterol. *Lavangadi Vati* prepared as per Ayurvedic formulary of India was found to contain 0.49 mg g^{-1} of eugenol, 0.07 mg g^{-1} of piperine and 0.007 mg^{-1} of ß-sitosterol. Marketed formulations (LV2 and LV3). LV2 contained 0.1105 mg g^{-1} of eugenol, 0.0679 mg g^{-1} of piperine and 0.00713 mg g^{-1} of ß-sitosterol, where as LV3 showed 0.1761 mg g^{-1} of eugenol, 0.066 mg g^{-1} of piperine and 0.006 mg g^{-1} of ß-sitosterol (Shailajan and Menon 2011).

Pre-clinical pharmacology

***Anti-tussive*:** An experimental study reported efficacy of *Lavangadi Vati* in the treatment of cough (Ojha et al. 1979).

Lodhrasava

Pre-clinical pharmacology

***CNS effects*:** *Lodhrasava* was evaluated for CNS effects in view of its observed sedative and tranquillsing effects in patients. It failed to elicit sedative, anti-psychotic, anti-convulsant, anti-depressant and anti-parkinsonism effects (Ravishankar and Sasikala 1986).

Clinical study

Dysfunctional uterine bleeding: *Pushyanuga churna* & *Lodhrasava* were selected for the study. Of total 46 cases studied, 12(29.26%) cases showed good response, 16 cases (39.02%) showed fair response, 9 cases (21.95%) showed poor response and 4 cases (9.75%) did not show any response after treatment. Five (10.86%) cases were dropped out from study. The treatment was found to be highly significant ($P < 0.001$) in reduction of uterine hemorrhage (Prameela Devi 2007).

Loknath Rasa

Standardization: *Loknath rasa* was prepared by different methods, i.e., *Gujaputa* and Electric muffle furnace. Muffle furnace method was found to be more effective than *Gujaputa*. On order to prevent adulteration, *Loknath rasa* was prepared separately by rubbing on juice of *Piper betel* and tap water. Iron, Aluminium and Calcium values were found higher in the water mardan while mercury, sulphar, magnesium and sulphate were found to be less in the different method of preparation.

Madhuhari Churna

Pre-clinical pharmacology

Antidiabetic: Hydroalcoholic extract of *Madhuhari churna* was administered orally at two dose levels 200 mg/kg and 400 mg/kg body weight, for 10 consecutive days in diabetic rats, and glibenclamide was kept as standard. Blood glucose (BG) and OGTT in normal healthy rats produced significant fall in BG and improved glucose tolerance.

Mahalaksadi Taila

Standardization: The specific gravity, saponification number, iodine number and acid number were 0.915, 199, 102 and 2 respectively. The drug resolved into two spots in the solvent n-butanol, acetic acid, water (75:15:10). The visible spectrum showed two peaks at 420-30 nm and 660 nm (Alam et al. 1989).

Mahalaxmivilas Rasa

Pre-clinical pharmacology

Genotoxicity: Comet assay were employed to study the endpoint of single/double-strand DNA breaks. The longer tail length observed was 54.6 ± 9.1 μm which was at 0.8% concentration of Diabecon and the tail lengths of *Mahalaxmivilas Rasa* were 37.3 ± 13.4 μm. The results revealed lack of induction of DNA damage as evidenced by the Comet assay, despite the presence of traces of transformed toxic heavy metals (Mishra and Mishra 2013).

Maharasnadi Quatha

Pre-clinical pharmacology

Anti-inflammatory and *analgesic*: *Maharasnadi Quatha* significantly and dose-dependently inhibited carrageenan-induced rat paw oedema (the inhibition at 3 h was greater than at 1 h after induction of oedema). *Maharasnadi Quatha* also increased the reaction time of rats in the hot-plate test (by 57% after the first hour of treatment) (Thabrew et al. 2003).

Clinical study

Rheumatoid arthritis: Twenty two patients suffering from rheumatoid arthritis were given *Vatari Guggulu* and *Maharasnadi Quatha* internally for 3 to 6 weeks. The results were assessed in terms of clinical recovery and functional improvement. Relief from pain, oedema and fever was statistically significant ($P < 0.01$) and fall in ESR levels was highly significant ($P < 0.001$) (Swamy and Bhattathiri 1998).

Mahamrutyunjaya Rasa

Pre-clinical pharmacology

Cardioprotective: Cardioprotective effect of *Mahamrutyunjaya rasa* laboratory prepared formulation (F1) and two marketed (F2 and F3) 25 and 50 mg/kg doses were studied in isoproterenol-induced myocardial infarcted rats. F1 and F2 did not affect the cell viability, while F3 decreased the cell viability in concentration and time-dependent manner (Rai and Rajput 2011).

Toxicity: Study shows that F1 and F2 possess cardioprotective property with higher safety, while formulation F3 cannot be used as cardioprotective due to its cytotoxic effects.

Mahayograj Guggulu

Pre-clinical pharmacology

Anti-inflammatory: *Mahayograj guggulu* showed dose-dependent anti-inflammatory activity with a maximum of 49% in paw edema, respectively, at a dose of 500 mg/kg. Prepared sample showed significantly better activity as compared to the commercial sample (Bagul et al. 2005).

Safety studies: A total of 40 Charles Foster strain albino rats of either sex with an average body weight of 160–250 g were divided into four groups (Groups I, II, III and IV), with 10 animals in each group. Group I served as the control, while Group II, III and IV rats received *Mahayograj guggulu* at a dose of 54 (dose equivalent to human therapeutic dose), 270 (five-times the dose equivalent to the human therapeutic dose) and 540 (10-times the dose equivalent to human therapeutic dose) mg/kg, p.o. for 120 days (Lavekar et al. 2010).

Mahayograj guggulu was found to be safe at all dose levels tested. No significant behavioral changes were noted in any of the groups studied. No major alterations were observed in hematology, serum biochemistry, necropsy and histopathology at the therapeutically advocated dose level.

Heavy metal estimation: Heavy metal concentrations were measured using an atomic absorption spectrophotometer. Heavy metal content measurement indicated levels of 25.8 µg/g for lead, 0.07 µg/g for mercury and 5.19 µg/g for arsenic.

Clinical study

Chorea: A study reported a case of eleven years old child who was treated with the combination of *Mahayograj Guggulu, Rasraj* and *Agnitundi Rasa* (Verma 1985).

Malla-Sindur

Pre-clinical pharmacology: Some pharmacalogial investigations have been done on *Malla-Sindur* in 1974–5 (Sawhney et al. 1974; Sawhney 1974–5).

Clinical study

Hemiplegia: In this study, *Malla-Sindur* was undertaken to see its efficacy in various symptoms of *Pakshaghata* (Hemiplegia). Statistically highly significant result was found in all symptoms. The overall effect of therapy shows that out of 30 patients, 1(3.33%) got good response, 10(33.33%) got fair response, 12(40%) got poor response and 7(23.33%) were in no response category (Naphade et al. 2011).

Manasamitra Vatakam

Pre-clinical pharmacology

Neuroprotective: Neuroprotective activity of *Manasamitra vatakam* was investigated on aluminium-induced neurotoxicity in rats. Following *Manasamitra vatakam* treatment, there was a significant recovery in the performance of the radial maze and muscle grip strength of the rats, as well as the levels of deoxyribonucleic acid, ribonucleic acid, and acetylcholinestrase in different parts of the rat's brain. The neuroprotective efficacy of *Manasamitra vatakam* was supported by histopathological observations (Thirunavukkaras et al. 2012).

Manikya Bhasma (Ruby calyx)

Standardisation: XRD Result of Raw Ruby showed Al_2O_3 as principal component. Mean Particle Size of Raw Ruby is 1071.58 nm. The particle sizes of *Hartaladi Manikya Bhasma* and *Manikya Bhasma* were respectively 43 nm and 63 nm (Wavare et al. 2014).

Manikya Pishti

Standardisation: X-Ray diffraction analysis of raw ruby showed Aluminum oxide as the principal component. Mean particle size of raw ruby was 107.58 nm and that of pishti was 52 nm (Wavare et al. 2014).

Mandura Bhasma

Physico-chemical characteristics: X-ray Diffraction analysis revealed that raw *Mandura* contained Fe_2SiO_4, and *Mandura bhasma* contained Fe_2O_3 and SiO_2. Scanning Electron Microscopy studies showed that the grains in *Mandura bhasma* were uniformly arranged in agglomerates of sizes 200–300 nm as compared to the raw *Mandura*, which showed a scattered arrangement of grains of sizes 10-2 microns (Mulik and Jha 2011).

Preclinical pharmacology

***Antibacterial*:** *Mandura bhasma* showed significant antibacterial activity against enteric bacterial pathogens such as *Escherichia coli, Staphylococcus aureus, Enterobacter aerogenes, Pseudomonas aeruginosa, Bacillus subtilis, Klebsiella pneumoniae, Salmonella typhi, Staphylococcus epidermidis, Salmonella typhimurium* and *Proteus vulgaris* using a disc diffusion method (Tambekar and Dahikar 2010).

Haematinic **and** ***cytoprotective*:** In Charles Foster strain rats of either sex, anemia was induced by administering mercuric chloride (9 mg/kg). *Lauha bhasma* and *Mandura bhasma* (11 mg/kg) were evaluated for their haematinic activity. The observed results suggest that *Lauha bhasma* and *Mandura bhasma* possess significant ($P < 0.05$) haematinic and cytoprotective activity (Sarkar et al. 2007).

***Hepatoprotective*:** The activities of acid lipase, alkaline lipase, lipoprotein lipase and hormone sensitive lipase exhibited significant alterations during CCl4 induced hepatic injury, indicating a role for these enzymes in the mobilization of fat from adipose tissue and accumulation of fat in liver and kidney. Simultaneous treatment with mandur bhasma prevented the paraffin mediated and CC14 mediated changes in the enzyme activities (Devarshi et al. 1986).

Mandura Parpati

Physicochemical analysis: Analytical study revealed that all the three parpatis which were used in the study had almost equal quantity of Ferric iron (30% to 35%) and Ferrous iron (4% to 5%). There was equal quantity of Hg and S in three *parpati*. XRD analysis of three samples of *Mandoor parpati* shows similarly major and minor phases present (Bavadekar et al. 2013).

Mandura Vataka

Preclinical pharmacology

***General pharmacology*:** Fine suspension of MV in 3% Tween 80 was used in all the experiments. *Mandura Vataka* is non toxic upto the dose of 2 g/kg (PO) in rats and mice. It had mild stimulating effect on CNS in rats and mice. It has also significant analgesic effect in rats and it shortened the duration of Diazepam induced narcosis in mice. It produced negative inotropic and chronotropic effects in isolated frog heart and rabbit atria. *Mandura Vataka* produced dose dependent and reversible antispasmodic effect against Ach and Barium chloride induced spasm in rat colon and rabbit jejunum (Ravishanker 1984).

Analgesic: *Mandura Vataka* (400 mg/kg) increased the threshold of tail flick response which was statistically highly that of pethidine (30 mg/kg) however the duration of action of *Mandura Vataka* is longer than that of pethidine (Ravishankar et al. 1986).

Manjisthadi Churna

Standardisation: A rapid, simple and accurate method with high performance thin layer chromatography has been developed to standardise *Manjisthadi churna* using rubiadin, sennoside and ellagic acid as markers. The rubiadin, sennoside-A and ellagic acid contents in *Manjisthadi churna* were found to be 0.014, 0.038 and 0.534% w/w, respectively (Patel and Patel 2013).

Manjishthadi Ghrita

Clinical study

Wound-healing: In a randomized control clinical trial, *Manjishthadi Ghrita* was evaluated clinically for wound-healing. It was used topically in postoperative wounds, mostly of ano-rectal cases, twice a day, for 21 days. Out of 45 patients, 24 patients in group A were treated with *Manjishthadi Ghrita* (treated group), while 21 patients in group B (standard group) were treated with povidine iodine ointment. Better result was observed in the treated group in comparison to the standard group (Baria et al. 2011).

Manjishtadi kashayam

Pre-clinical pharmacology

Antioxidant: Among five decoctions, *Manjishtadi kashayam* exhibited higher antioxidant properties (total phenolic content-15.61 ± 0.006 mg/g wt, EC50-7.2 µg/ml).

Mayurapuccha Bhasma (Calx of peacock feather)

Physico-chemical analysis: Bhasmas (one by burning on ghee flame and the other by giving Gajapuṭa (burning the peacock feathers at about 1000°C by using a thousand cow dung cakes)) exhibited marked difference in color, moisture content, and percentage of inorganic compounds. The Bhasma prepared by *Gajapuṭa* method contains essential and beneficial inorganic elements, electrolytes in larger quantity, and lower moisture content (Kotrannavar et al. 2012).

Mehamudgara Vati

Clinical study

Type 2 diabetes mellitus: In Group A, *Mehamudgara vati* was given 3 tab thrice a day with lukewarm water for three months and in Group B, the patients who were already taking modern antidiabetic treatment, although their blood sugar level was not well under control, were additionally given *Mehamudgara vati* in the same manner. The formulation has shown a highly significant decrease in the fasting and post-prandial blood sugar level (Tanna et al. 2011).

Mrga Srnga Bhasma

Toxicology: No toxic effects were observed in a group of five male rabbits (Srivastava 1992).

Mukta Bhasma (Pearl calx)

Preclinical pharmacology

***Gastroprotective*:** It was observed that *Mukta bhasma* produced significant ($P < 0.001$) 57.53% and 59.18% protection in cold restraint stress induced gastric ulcer and 54.19% and 61.39% protection in Diclofenac induced ulcer at 60 mg/Kg and 120 mg/Kg dose levels respectively, when compared with control. TBARS of stomach in ulcer induced rat was also reduced by *Mukta bhasma* (Dubey et al. 2009).

Chemical Analysis: Calcium carbonate.

Muktashukti Bhasma (Pearl oyster calx)

Chemical Analysis: Calcium carbonate, calcium phosphate, aluminium oxide, magnesium oxide and organic matter (Sane et al. 1983).

Standardization: Five batches of *Mukta Shukti Bhasma* were prepared and standardization was attempted by maintaining batch manufacturing records of individual batches. During pharmaceutical procedures like *Shodhana, Bhavana, Marana*, etc. due care of temperature, its duration, percentage of weight gain or loss and the cost factor of the end product, etc. were considered. The average weight loss observed was 14.62 g, i.e., 2.92% during *Mukta Shukiti Shodhana*. Average weight loss found was 3835.24 g, i.e., 7.05% (Kumar, Parmar and Patgiri 2012).

Preclinical pharmacology

***Anti-inflammatory*:** *Muktashukti bhasma* inhibited acute and subacute inflammation in albino rats as induced by subplanter injection of carrageenan, histamine, 5-HT, nystatin and subcutaneous implant of cotton pellets. In all the test procedures the antiinflammatory response of 1000 mg/kg *Muktashukti bhasma* was comparable to the response observed with 300 mg/kg acetylsalicylic acid (Chauhan et al. 1998).

***Gastroprotective*:** In a study carried out on two hundred and thirty-eight inbred Albino rats, *Muktashukti Bhasma* caused reduction in ulcer score and ulcer index with all the doses. *Muktashukti Bhasma* caused significant reduction ($p < 0.001$) in free and total acidity, and acid output only at 300 and 1,000 mg/kg dose. *Muktashukti Bhasma* and

Ranitidine caused decrease in peptic activity at all dose levels. *Muktashukti Bhasma* also raised pH (Chouhan et al. 2010).

Mustakarishta

Three different methods were performed for the preparation:

1. Traditional earthen pot method
2. Steel container method
3. Wooden container method

All the formulations were analyzed for various qualitative and quantitative parameters according to WHO guidelines and the results were compared with the marketed formulation. With the change in method of preparation, considerable variations were observed in the parameters and the results of ME were found to be most significant (Kadam et al. 2012).

Naga Bhasma (Incinerated Lead or Lead Calyx)

Elemental composition: Depending on the preparation, the lead content varies from 14.5% to 68.14%. The lead content of *sastiputa naga bhasma* reported by Nagaraj was 64.24% (Nagarsi 1992) whereas in their sample of the drug, Keen et al. reported a lead content of 19%.

Physico-chemical characteristics: Thermogravimetry analysis showed that *naga bhasma* sample was thermally stable until 900°C, indicating the absence of free organic molecules. The FTIR spectra revealed that all the samples contained organic moieties probably in the form of a complex. Particle size and surface area analysis of the *naga bhasma* samples indicated the presence of micron-sized particles. Elemental analysis indicated the presence of arsenic impurity in the samples. Electron microscopy studies revealed that *bhasma* contained particles in micron & sub-micron ranges. Energy dispersive X-ray Analysis too showed the presence of arsenic along with Lead. XRD showed the lead oxide phase in all three samples (Nagarajan et al. 2012).

Characterisation of *Naga Bhasma*: Purification process removes heavy metals other than lead, apart from making it soft and amenable for trituration. The use of powders of tamarind bark and peepal bark maintains the oxidation state of lead in Jarita Naga (lead oxide) as Pb(2+). The repeated calcination steps result in the formation of nano-crystalline lead sulphide, the main chemical species present in *Naga bhasma* (Nagarajan et al. 2014).

Standardisation: Results of the present study on infrared, Raman and X-ray photoelectron spectroscopy along with thermal measurements identify the presence of carbonaceous material in the drug along with other compounds. In addition, this work also suggests the science and mechanism behind such complex preparations which could help in standardization of such medicines (Singh and Rai 2012).

The present study was planned to standardize *Naga bhasma* prepared by using Vasa as herbal media. Prepared *Naga bhasma* was subjected to tests mentioned in

Ayurvedic texts and physico-chemical analysis such as pH Value, Total Ash, Loss on drying and acid insoluble ash. Twenty eight days are required to prepare *Naga bhasma* with 3.76% weight loss (Rajput et al. 2013).

Preclinical pharmacology

***Testicular Regenerative* Potential:** It was observed that the test drug when given simultaneously with Cdcl2 showed marked prevention of toxic effects of Cdcl2 and when given alone after 36 hours of Cdcl2 administration, showed a notable regenerative potential on partially degenerated testes. It has showed specific regenerative effect on germinal epithelium of testes (Singh et al. 1989).

***Diabetes mellitus*:** There are 44 formulations of *Nāga bhasma* mainly indicated for polyuria. Haridrā (*Curcuma longa* Linn.), Āmalakī (*Emblica officinalis*), Guḍūci (*Tinospora cordifolia*) and Madhu (honey) enhance the antidiabetic action of *Nāga bhasma* and also help to prevent diabetic complications as well as any untoward effects of *Nāga bhasma* (Rajput et al. 2013).

Toxicity: *Naga bhasma* contains lead in nano-crystalline (~60 nm) lead sulfide form (Pb(2+)) associated with the organic contents and different nutrient elements coming from the herbs used during the preparation. *Naga bhasma* prepared was found to be totally safe in histopathology study on rats at a dose of 6 mg/100 g/day (Singh et al. 2010).

Acute and sub-acute toxicity: Experimental studies showed that Pb, PbO, PbSO4, and PbS preparations available in the market are toxic. X-Ray diffraction analysis proved that *Sastiputa naga bhasma* is chemically PbS; and reported to be non-toxic in experimental and clinical studies (Nagarsi 1992).

The test drug at the dose of 10 mg/kg studied does not produce any significant degenerative changes. From toxicological point of view the test drug does not produce any serious tissue damage at therapeutic dose and intermediate dose level. On histopathological study, in highest dose the stages of spermatogenesis was reduced, the structure of graffian follicles showed separation in their wall, the stages of spermatocytes looked degenerative (Dash et al. 2012).

A study reported the case of a 58-year-old woman who presented with abdominal pain, anemia, liver function abnormalities, and an elevated blood lead level. The patient was found to have been taking the Jambrulin prior to presentation. Chemical analysis of the medication showed high levels of lead (Gunturu et al. 2011).

Narcha Churna

Standardisation: Physico-chemical standardization of *Narcha churna* has been discussed in a study (Agarwal and Girija 2005).

Nardiyalaxmi Vilas Ras Mishran

Pre-clinical pharmacology

Anti-diuretic: *Nardiyalaxmi Vilas Ras Mishran* does not affect the urinary function at lower doses but at higher doses it revealed anti-diuretic activity in albino rats (Maurya et al. 1982).

Navbal Rasayan

Toxicity: Oral administration of graded doses of *Navbal Rasayan* (upto 3.0 g/kg) did not produce any acute or chronic toxicity in rats. Oral pretreatment of guinea pigs with *Navbal Rasayan* (1.5 g/kg for 3 days) increased 36.43 time the histamine dose while the agonistic effect of acetylcholine and 5-hydroxytryptamine was completely attenuated. Further, NR neither exhibited any analgesic, sedative effect or anticonvulsant effect in rodents (Chandra and Mandal 2000).

Navratna Rasa

Toxicity: *Navratna rasa* was screened for its safety/toxicity studies in acute and chronic models. The chronic toxicity study reveals that, the test drug has no serious toxicity potential to most of the important organs in therapeutics doses (Lavekar et al. 2009).

Nava-Yasa Churna

Pre-clinical pharmacology

Hepatoprotective: The enzyme levels of GPT, GOT and ALP were significantly elevated in CCl4 treated rats in comparison to control group. Pretreatment with *Nava-Yasa churna* significantly ($P < 0.01$) reduced the raised enzyme levels (Shastry et al. 2013).

Nimbadi Taila

Pre-clinical pharmacology

Anti-inflammatory: *Nimbadi Taila* was investigated for activity in Swiss albino rats against Piroxicam gel as standard reference and normal saline as control by Randall and Baroth method. The time to achieve reduction of inflammation in the rat paw was determined. The topical application of *Nimbadi Taila* exhibited significant anti-inflammatory activity when compared with the Piroxicam gel and normal saline (Yamini and Chalapathi 1982).

Clinical study

Dermatitis: A study reported efficacy of *Nimadi Taila* and *Somraj Churna* in the treatment of chronic dermatitis (Shaw et al. 1982).

Nidigdhikadi Kvatha

Bronchial asthma: In this study, 40 patients of *bronchial asthma* in remission stage were registered, out of which 20 patients were included in the trial group and treated with *Nidigdhikadi kvatha* & another 20 patients were included in control group and treated with tablet of Deriphylline-retard by simple random sampling method. In both the groups the effect of drugs on symptoms of *bronchial asthma* was significant; however it was observed that the trial drug formulation proved to be effective in relieving coughing and difficult expectoration than the control drug (Chavan et al. 2013).

Nirgundi Ghana Vati

Clinical study

Rheumatoid arthritis: Total 50 randomly selected patients of *Amavata* were registered; among them 45 completed the treatment. *Kshara Basti* in the format of *Kala Basti* was given to these patients and *Nirgundi Ghana Vati* was given for one month. Statistically significant improvement was found in ESR, RA factor (quantitative) and also highly significant results were found in symptoms of *Amavata*. Moderate improvement was seen in 40% of patients, 35.56% patients got marked improvement, while mild improvement was found in 24.44% of patients (Thanki et al. 2012).

Nisamalaki Churna

Pre-clinical pharmacology

Antidiabetic: Alloxan was administered as a single dose (120 mg/kg, b.wt.) to induce diabetes. There was administration of different formulations suspension of *Nisamalaki Churna* tablet (200 mg/kg) for 14 days, to alloxan-induced diabetic rats. The fasting blood sugar levels in alloxan-induced diabetic rats were investigated. Oral administration of different formulations suspension *Nisamalaki Churna* tablet (200 mg/kg) for 14 days exhibited a significant reduction in blood glucose in alloxan induced diabetic rats as compared with the standard drug Glibenclamide (5 mg/kg) (Kumar et al. 2012).

Nishadi Taila

Clinical study

***Fistula-in-ano*:** A clinical trial was conducted upon 30 patients of Bhagandara with Ayurvedic intervention of *Apamarga kshara sutra* along with *nishadi tail purana* in one group and *Apamarga kshara sutra* alone in another group. Group with *Apamarga kshara sutra* along with *nishadi tail purana* showed more improvement (Mallappanavar 2013).

Nityananda rasa

Clinical study

***Filariasis*:** A field clinical trial of *Nityananda rasa* in the treatment of *filariasis* has been reported (Mishra et al. 1979).

Nyagrodhadi Churna

Standardisation: The obtained values of physical and chemical parameters in this study can be used to lay down pharmacopoeial standards to be followed for traditional preparation of *Nyagrodhadi churna* with batch-to-batch consistency (Gopala Simha and Laxminarayana 2007).

Padmapatradi Yoga

Clinical study

***Bronchial asthma*:** A study was carried out in 40 patients of either sex in between the age of 15–65 years to assure the clinical response of *Padmapatradi yoga* in bronchial asthma. *Padmapatradi yoga* was effective in increased peak expiratory flow rate, breath holding time, and reduced the absolute eosinophil count of studied cases and was also found statistically highly significant at $p < 0.001$ level.

Palasha Kshara

Clinical study

***Fistula-in-ano*:** A comparative study proved that *Palasha Kshara Sutra* is equally good as *Apamarga Kshar Sutra* with marked reduction of symptoms like pain, burning sensation, irritation and other skin reactions.

***Urolithiasis*:** The study was carried out on 50 cases of urolithiasis of which 24 cases were of renal calculi and 26 cases were of ureteric calculi. At the end of the study it was concluded that different diet factors, low urinary output and suppression of urge to avoid urine play major role in the formation of calculi. The effect of the drug was found more marked in the expulsion of the ureteric calculi as compared to renal calculi (Kumar and Kumar 1995).

Palasha Pushpadi Churna

Clinical study

Diabetes mellitus: In a clinical trial, *Palasha Pushpadi Churna* was administered in the dose of 5 gm thrice a day with lukewarm water for a period of two months. Significant improvement in the clinical symptoms was found in the study. Fasting blood sugar level was decreased by 30.69% and post prandial blood sugar level was decreased by 25.15%. A decreased of 72.37% was also noted in the urinary sugar level.

Panchakola Avaleha

Animals were administered three doses of *Panchakola Avaleha* by oral routes, viz. higher (500 mg/kg/day), middle (250 mg/kg/day), and therapeutic dose (50 mg/kg/day) for 28 consecutive days. Effects of the test drug on hematological, biochemical, and histopathologic parameters were evaluated. This study revealed normal behavior, no mortality, and no significant changes in hematological, biochemical, and histopathological examinations (Singh et al. 2012).

Panchamrita Parpati

Clinical study

Chronic colitis: A clinical study reported efficacy of *Panchamrita parpati* in the treatment of chronic colitis (Deshpande et al. 1977).

Ulcerative colitis: A clinical study reported efficacy of *Panchamrita parpati* in the treatment of ulcerative colitis (Sharma et al. 2006).

Pancaguna Taila

Physicochemical characters: Thermodynamical parameters show that *Pancaguna taila* has exothermic reaction, which upon applying mechanical heat helps the *Pancaguna taila* in permeating the skin and thus has direct action on the diseased part (Saxena et al. 1990).

Pancha Nimba Churna Vati

Clinical study

Eczema: Role of jalaukavacharan and *Pancha Nimba Churna Vati* has been highlighted in treatment of *Eczema* (Patankar et al. 2005b).

Panchtikta Ghritha

Pre-clinical pharmacology

Anti-inflammatory **and** ***analgesic***: The *Avartita Panchtikta Ghritha* significantly inhibited carrageenan induced paw oedema and formalin induced paw licking response. However it was failed to suppress the formalin induced paw oedema, whereas *Murchita Panchtikta Ghritha* significantly inhibited formalin induced pain response but failed to suppress carrageenan induced paw oedema and formalin induced paw oedema.

Panchatikta Ghrita Guggulu

Clinical study

Osteoarthritis: In this study, a total 49 patients having the complaints of osteoarthritis were randomly divided into two groups. In Group A, patients were treated with *Panchatikta Ghrita Guggulu Vati* along with *Abhyanga* and *Nadi Swedana* and in group B patients were treated with only *Abhyanga* and *Nadi Swedana*. The data shows that *Panchatikta Ghrita Guggulu* along with local *Abhyanga* and *Nadi Swedana*, i.e., group A has provided better relief in the disease osteoarthritis (Akhtar et al. 2010).

Panchvalkala

Pre-clinical pharmacology

Antioxidant: *Panchvalkala* and its individual components showed significant antiradical activity by bleaching 1,1-diphenyl-2-picrylhydrazyl radical (EC_{50} ranging from 7.27 to 12.08 µg) which was comparable to pyrogallol (EC_{50} 4.85 µg). All the samples showed good superoxide scavenging potential (EC 50 ranging from 41.55 to 73.56 µg) comparable to ascorbic acid (EC 50 45.39 µg) in a dose-dependent manner (Anandjiwala et al. 2008).

Panchavaktra Rasa

Standardisation: The alkaloid content of different parts of the drug *Panchavaktra rasa* fingerprint profile has been determined for the three experiments in the different times. The UV spectrum of the common peaks with large abundance appeared at Rf 0.46, 0.33, 0.28, 0.30, 0.41, 0.49, 0.56, 0.57, 0.59, 0.99 were found to be superimposable revealing the presence of the same constituents in all the three samples (Bandari et al. 2011).

Clinical study

Rheumatoid arthritis: 50 patients were selected and a trial drug was advocated in a dose of 300 mg (2 tablets) of *Panchavaktra rasa* twice a day with *Trikatu* and *Arka moola twak kashaya* as *anupana*. The relief of pain, stiffness, dyspepsia, fatigue, constipation were found highly significant (P < 0.001) and same results similar results in reduction in inflammation, ESR levels and RA Factor (Bandari et al. 2012).

Pancasama Churna

Standardisation: The loss on drying at 105°C in Pancasama churna was found to be 8.30%. Total ash value of plant material indicated the amount of minerals and earthy materials present in the plant material. Analytical results showed total ash value of 23.50%. The amount of acid-insoluble siliceous matter present in the plant was 3.52%. The water-soluble extractive value indicated the presence of sugar, acids and inorganic compounds.

Pashanabhedadi Ghrita

Pre-clinical pharmacology

Anti-urolithiatic: The calculi were induced by a gentamicin injection and ammonium oxalate rich diet. Test drug was administered concomitantly in the dose of 900 mg/kg for 15 consecutive days. Rats were sacrificed on the 16(th) day. Concomitant treatment of *Pashanabhedadi Ghrita* attenuated blood biochemical parameters non-significantly, whereas it significantly attenuated lipid peroxidation and enhanced glutathione and glutathione peroxidase activities (Gupta et al. 2012).

Phalaghrita

Clinical study

Infertility: The present study was carried out in a sample of 30 numbers of female infertility cases with the use of *phalagritam* in the form of intra uterine insufflation. After the treatment of three consecutive cycles, an overall encouraging result was observed (Otta and Tripaty 2012).

Pinda Taila

Standardisation: The analytical values can beseem as preliminary reference standards; these values are mostly related to the purity of the sesame oil. The thin layer chromatography studies of the *Taila* are more useful to detect the presence of single drugs present in *Pinda Taila* (Hepsibah et al. 1996).

Pippalyadi Yoga

Pharmacognostical and phyto-chemical evaluation: Phyto-chemical analysis showed Loss on drying 10.07% w/w, ash value 6.55% w/w, water soluble extract 14% w/w, methanol soluble 13.40% w/w, particle consistency above 60 mesh 4.10% w/w, between 60–85 mesh 9.20% w/w, between 85–120 mesh 13.30% w/w & below 120 mesh 73.37% w/w & pH 5.0 (Pandya et al. 2014).

Toxicity: Developmental toxicity of *Pippaliyadi yoga* was studied by administering three doses, viz. 140, 300 and 700 mg/(kg day) to gravid females from day 6 to day 16 of gestation. Pippaliyadi did not have any adverse developmental effects with low doses, however, with the five times higher dose, a decrease in body weight of the pups was observed. The reproductive performance of the progeny born to mothers treated with pippaliyadi was not significantly affected (Nafisa et al. 2007).

Clinical study

Infertility: Eleven patients presenting with complaint of failure to conceive for minimum one year with two consecutive anovulatory cycles in serial trans-vaginal sonography were included in the study. Out of them 10 completed the treatment. *Pippalyadi Yoga* was given in the dose of 4 gm before meal orally twice a day for two months after stoppage of menstruation. The statistical significance of results on selected criteria established that *Pippalyadi Yoga* mentioned in the management of infertility is effective treatment modality in anovulation (Pandya S and Dhiman 2014).

Pravala Bhasma (Coral calx)

Physicochemical characterization: Fourier transform infrared (FTIR) spectroscopy, bands appearing in final product spectra showed a significant shift in infrared vibration frequency as well as intensity when compared with the raw material, which was indicative of formulation of *bhasma*. X-ray diffraction revealed that raw material contained CaCO3 whereas in case of final product of *bhasma*, CaO was identified. Scanning electron microscopy revealed the difference in particles size of *bhasma* (10–15 μm) and raw material (100–150 μm). Energy dispersive X-ray analysis showed presence of different concentration of carbon in both the samples (Mishra et al. 2014).

Chemical Analysis: Qualitative analysis indicates presence of Na, K, Ca, Mg, Sr, Zn, Cu, Cd, Mn, Co, Sn, Mo and carbonate (Dixit and Shivhare 1988). Calcium is present in Praval bhasma as calcium oxide. Calcium as CaO% w/w is 72.067% (Saini et al. 2013).

Standardization: A study has been conducted to standardize the temperature for the preparation of *Pravala Bhasma* by using Electric Muffle furnace (Dixit and Rao 1998).

Preclinical pharmacology

A study was done to compare the oral bioavailability of *Pravala moola bhasma, Pravala shakha bhasma, Pravala moola pisti* and *Pravala shakha pisti* in dogs aging from 5 to 10 months. The single oral administration of the drug, two treatments, = and randomised design test methodology were applied. Statistically significant difference was observed in Tmax of *Pravala moola bhasma* and *Pravala moola pishti* as well as *Pravala shakha bhasma* and *Pravala shakha pishti*. It can be concluded that different preparations of *Pravala* have different mode of absorption in monogastric animals (Kaushal et al. 2010).

Bone mineralization: In the present study, twenty-four female rats were ovariectomized, 12 sham operated, divided into three groups of 12 each, fed on a low calcium diet (0.04% Ca) and treated either with vehicle or *Praval bhasma* (65 mg/kg body weight, twice a day) for 16 weeks. Compared to sham rats, concurrent calcium deficiency animals showed an increase in urinary excretion of calcium and phosphorus, decreased femoral weight and density which were significantly reversed in *Praval bhasma* treated animals.

Measurement of cortical bone morphometric indices by CT-scanning technique showed increased medullary width and cross-sectional area, decreased periosteal area, combined cortical thickness and cortical area/periosteal area in concurrent calcium deficiency animals compared to sham and PB-treated group. Ash weight, percent ash, ash Ca and ash P levels were lower in concurrent calcium deficiency animals than in sham or *Praval bhasma* group (Reddy et al. 2003).

Clinical study

Hyperacidity: In this study, two samples of *Pravala Mula Bhasma Sarjika Kshara* [PM(S)B] and *Tandulikya Swarasa Shodhita* [PM(T)B] and two samples of *Pravala Shakha Bhasma Sarjika Kshara* [PS(S)B] were prepared according to the authentic process. These all four *Bhasma* were studied to evaluate the efficacy of the drugs in the patients of hyperacidity. Results of the study suggest that the effect of *Pravala Shakha Sarjika Kshara Shodhita Bhasma* [PS(S)B] was better than all three on different clinical parameters of the patients (Tiwari et al. 2008).

Important Note: *Pravala Mula* and *Pravala Shakha* are two different species. *Pravala Mula* is *Tubipora musica* and *Pravala Shakha* is *Corallium rubrum*.

Puga Khanda

Physico-Chemical profile: *Puga Khanda* was prepared in three batches as per the classical reference. All the samples had 60% of sugar needed for preservation and 2/3rd of it was non reducing sugar. The total alkaloids ranged from 0.002 to 0.004% w/w. In TLC study, all the samples showed a similar pattern except the 2nd sample of *Puga Khanda* (Baragi et al. 2014).

Punarnavasava

Chemical and Biological evaluation: Two batches of three brands of *Punarnavasava* (A1, A2, B1, B2 and C1, C2) were purchased from market. The concentration of eupalitin 3-O-ß-Dgalactopyranoside was different in all three brands but batch to batch variation was not very high. The antiarthritic activity of *Punarnavasava* and *B. diffusa* extract at a dose of 100 mg/kg was comparable to that of diclofenac (Kasture and Gharate 2013).

Punarnavashtak Kwath

Chemical constituents: *Punarnavashtak Kwath* contains berberine and gallic acid at 0.08 and 4.9%, respectively.

Preclinical pharmacology

***Hepatoprotective*:** Administration of *Punarnavashtak kwath* produced a significant hepatoprotective effect as demonstrated by decreased levels of serum liver marker enzymes, and serum bilirubin and an increase in protein level. Thiopentone-induced sleeping time was also decreased in the *Punarnavashtak kwath*-treated animals compared with the CCl(4)-treated group (Shah et al. 2011).

Clinical study: *Punarnavashtak Kwath* was standardized by High performance thin layer chromatography using gallic acid and berberine as biomarkers. *Punarnavashtak Kwath* was given in the dose of 20 ml kwath daily for eight weeks. The results showed significant changes in liver functions tests. There was no report of adverse effects recorded (Shah et al. 2010).

Pushyanuga Churna

Clinical study

***Leucorrhoea*:** A study reported efficacy of *Pushyanuga Churna* in the treatment of leucorrhoea (Geetha et al. 1991).

***Dysfunctional uterine bleeding*:** *Pushyanuga churna* & *Lodhrasava* were selected for the study. Of total 46 cases studied, 12(29.26%) cases showed good response, 16 cases (39.02%) showed fair response, 9 cases (21.95%) showed poor response and 4 cases (9.75%) did not show any response after treatment. Five (10.86%) cases were dropped from study. The treatment was found to be highly significant (P < 0.001) in reduction of uterine hemorrhage (Prameela Devi 2007).

Puskara Mooladi Churna

Clinical study

***Bronchial Asthma*:** Six weeks treatment with *Puskara Mooladi churna* showed a significant effect of increase in Pulmonary Function values. The mean grade score plus standard deviation before trial of FEV, FVC, and PEFR were 62.6 ± 15.06, 2.03 ± 0.53 and 189 ± 44.05 respectively. After six weeks of treatment with *Puskara mooladi churna* FEV, FVC and PEFR showed highly significant results with values 63.45 ± 15.9, 2.81 ± 0.33 and 199.6 ± 41.58 respectively (Sai Prasad et al. 2010).

Pushpadhanwa Rasa

Pre-clinical pharmacology

Spermatogenic: Oral administration of *Pushpadhanwa Rasa*-A, B, C was given in 10 mg/kg dose along with *Solanum xanthocarpum* methanol extract solution for 60 days. Three samples of *Pushpadhanwa Rasa* were prepared to assess their spermatogenic effect. *Pushpadhanwa Rasa*-A, B, C show significant result. The spermatogenic effect was further supported by reversal of *S. xanthocarpum* methanol extract induced histopathological changes. *Pushpadhanwa Rasa*-B proved better efficacy than the other two. Overall analysis of the histopathological observation indicates that *Pushpadhanwa rasa*-C has the potential of producing moderate toxicity in liver, heart and kidney on long-term administration.

Note:
(*Pushpadhnwa rasa*-A): Prepared traditionally with herbal triturating drugs
(*Pushpadhnwa rasa*-B): Prepared with extract of herbal triturating drugs
(*Pushpadhnwa rasa*-C): Prepared without herbal constituents.

Clinical study

Polycystic Ovarian Syndrome: A total no. of 12 females with PCOS were selected and were divided equally into two groups. The group A were treated with *Pushpadhanwa Rasa* prepared traditionally as mentioned in classics where as in group B *Pushpadhanwa Rasa* prepared by using herbal extract. The treatment was given for two months at the dose of 250 mg. On starting *Pushpadhanwa Rasa* therapy a significant symptomatic relief of pain and swelling in the lower abdomen was reported by patients. There was a significant ($P < 0.001$) reduction in the size of the cyst at the end of the therapy. Dissapearance of cyst was noted in nine patients at the end of the 60-day study period.

Rajata Sindura

Standardization: Scanning electron microscope coupled with energy dispersive spectroscopy analysis has shown mercury 86.21%, sulfur 13.27% as major elements; iron, calcium, potassium, magnesium and silver were other detected minor elements. X-ray diffraction report revealed the chemical nature of *Rajata Sindura* as HgS compound as having cubic a crystal structure (Gokarn et al. 2014).

Raktapittantaka Louha

Toxicology: *Raktapittantaka Louha* was administered orally to the male albino rat. After 46 days of treatment period, animals were fasted for 18 hours after the last administration. Biochemical studies were conducted. Triglyceride level was significantly ($p = 0.001$) lowered in the RPT group than the corresponding control group. Total cholesterols, LDL, VLDL and HDL levels were raised significantly ($p =$

0.05) in the RPT group. The amount of Albumin, Creatinine and Urea in serum were significantly (p = 0.05) increased in RPT group (Mamun Al-Amin and Kamal 2013).

Rasa Bhasma

Acute toxicity: In fifteen Wistar Rats after seven days of exposure to *Rasa Bhasma* with recommended therapeutic dose, no significant changes in haematological parameters & clinical chemistry parameters were found. No pre-terminal deaths occurred in rats, which received the test compound at therapeutic dose levels. No abnormalities in physical, physiological, clinical chemistry, hematological parameters and no gross necropsy changes were observed on administration (oral) of test compound prepared according to the classical literature (Prasad et al. 2013).

Rasaka Bhasma

Physicochemical analysis: After using classical parameters for the identification of *Rasaka*, two samples were taken up for analysis: ZnCO3 and ZnO. Both the samples were subjected to *Shodhana* using *Nimbu Swarasa*. A fragrant odor in the ZnO sample after Shodhana was obtained, which may have been due to trituration with *Nimbu Swarasa*. The prepared Bhasmas were subjected to qualitative and quantitative analysis. The pH of zinc carbonate before *Shodhana* was 8.73, after *Shodhana* 7.79, and after *Marana* 7.73. In the case of zinc oxide, the pH was 6.04 before *Shodhana*, 6.48 after *Shodhana*, and 5.9 after *Marana* (Shubha and Hiremath 2010).

Pre-clinical pharmacology

Antimicrobial: Antimicrobial result of *Rasaka bhasma* prepared out of $ZnCO_3$ has shown a better result than *Rasaka bhasma* prepared out of ZnO. *Bhasma* prepared out of ZnCO3 has shown better antimicrobial results over *Str. pyogenes* compared to other organisms at the concentration of 50 µl (Shubha and Hiremath 2010a).

Rasa Karpura (Per-chloride of mercury)

Standardisation: Preliminary studies were undertaken for pharmaceutical standardisation of *Rasa karpura* (Prasad et al. 1992; Joshi and Prabhakara Rao 1992).

Preclinical pharmacology

Antibacterial: Preliminary studies reported antibiotic properties of *Rasa karpura* (Choudhary et al. 1999).

Toxicology: Dermal toxicity of *Rasakarpura Drava* (solution form of *Rasa karpura*) was carried out in albino rats and compared with the solution of mercuric chloride. Results show that *Rasa karpura* is relatively safe on sub-acute dermal application in terms of haematological, serum biochemical and histopathological parameters, while mercuric chloride is apparently toxic (Mehta et al. 2012).

In the male as well as female group, there were statistically very highly significant increases in both the total protein and albumin content in the plasma. Statistically very highly significant decrease in lipid profile was observed in both sexes of the experimental animals. Considerable decrease in the low density lipoprotein and high density lipoprotein were noted. There was an increase in the uric acid content in the plasma of male rats but this was not significant. Similar result was also observed in female rats (Abdullah-Al-Mamun et al. 2014).

Clinical study

***Skin diseases*:** The clinical study was carried out on 80 patients of *Kshudra Kustha* by using *Rasa karpuradrava* (0.1% solution in distilled water) and the same was compared with a standard drug, i.e., *Gandhaka Malahara*. The obtained results of the clinical trial suggest that both drugs have a highly significant result on the cardinal symptoms of *Kshudra Kustha* (Mehta et al. 2011).

Rasamanikya (Arsenic trisulfide)

Chemistry: X-ray diffraction implies that *Rasamanikya* is a mixture of arsenic trisulfide and arsenic pentaoxide (Pandey et al. 1999).

Analytical evaluation: In the present study, the samples viz. ashudha *haratala*, *shodhita haratala*, *Rasamanikya-Abhraka* (mica sheet) method, *Rasamanikya-* (modified) *sharava* samputa method and *Rasamanikya*-fuse bulb method were subjected for X-ray diffraction and ICP-AES apart from the ancient parameters. It was found that were minimal difference found in the three samples of *Rasamanikya* but when the three samples were compared together it was assessed that *Rasamanikya*-I prepared out of *Abhraka Patra* method showed better results from the standardization point of view and quality assurance of *Rasamanikya* (Sud et al. 2011).

Preclinical pharmacology

***Antibacterial*:** An animal study reported antibacterial efficacy of *Rasamanikya* on *Staphylococcus aureus* and *Pseudomonas aeruginosa* (Shalini et al. 2011).

Clinical study

***Bronchial asthma*:** A compound formulation based on *Rasamanikya* was tried in 26 patients of *bronchial asthma*. The drug is found to be effective irrespective of age, sex and duration of illness. Severity and frequency of the attack is minimised after the administration of the drug. It also reduced the eosinophilic count significantly (Shankara et al. 1978).

***Rheumatoid arthritis*:** A clinical study reported efficacy of *Rasamanikya* in rheumatoid arthritis (Dash and Das 1994).

Genotoxicity: In the present study, there was no incidence of genotoxicity induced by this *Ras manikya ras* in the Micronucleus assay or the Comet assay. The ready absorption of arsenic in the gut and its excretion in the urine might have contributed to this effect (Sathya et al. 2008).

Toxicity: The three samples when compared together showed that *Rasamanikya* prepared out of classical *Abhraka Patra* method and modified *Sharava Samputa* method showed minimal histopathological changes proving its non-toxicity, whereas *Rasamanikya* prepared out of electric bulb method showed mild toxicity, but with chances of recovery. Acute toxicity study showed no immediate and evident toxic signs and mortality in histopathology reports and liver function test. However, sub-acute toxicity study showed mild to moderate fatty changes in liver (Sud et al. 2013).

Rasa Parpati

Standardization: During the study, it was observed that the time between the stages (*pakas*) are very short; the sulphur is in more free form in the *Kajjali* as compared to *parpati*. No bile salts/bile pigments and chlorophyll were transferred from cow dung and banana leaves respectively (Rao and Vaish 2014).

Clinical study

Ulcerative colitis: A clinical study was done on 60 patients, selected randomly and divided in two groups. Group A, i.e., trial group patients were treated with oral dose of *Rasa Parpati* & *Jatyadi Ghrita Matra Basti*. The Group B patients, i.e., control group, were treated with Salazopyrine. The clinical assessment was done on the basis of clinical presentation of ulcerative colitis as well as colonoscopic findings, before and after the treatment (Shekokar et al. 2013).

Rasa Pushpa

The group which received *Rasa pushpa* in a dose of 12 mg/kg showed highly significant anti-inflammatory activity in acute inflammation against the standard group which received diclofenac. Effect of *Rasa pushpa* was found to be dose dependent. In sub-acute inflammation, decrease in weight of granuloma by *Rasa pushpa* proved its significant proliferative activity (Shrivastava et al. 2013).

Rasraj Rasa

Clinical study

Chorea: A study reported a case of eleven years old child who was treated with the combination of *Mahayograj Guggulu, Rasraj* and *Agnitundi Rasa* (Verma 1985).

Rasa-Sindoor (Sub-chloride of mercury)
Pre-clinical pharmacology

Neuroprotective: A study investigated neuroprotective effects of *Amalaki Rasayana* and *Rasa-Sindoor* in fly models of polyQ (127Q and Huntington's) and Alzheimer's disorders. Both formulations prevented accumulation of inclusion bodies and heat shock proteins, suppressed apoptosis, elevated the levels of heterogeneous nuclear ribonucleoproteins and cAMP response element binding protein and at the same time improved the ubiquitin–proteasomal system for better protein clearance in affected cells (Dwivedi et al. 2013).

Toxicity: Total 30 albino rats were taken for the study and they were divided into five groups. The first group was control group while other the four groups were administered two samples of *Hingulottha Parada Rasa-sindoor* and *Shodhita Parada Rasa-sindoor* in two different doses (50 mg/kg and 100 mg/kg respectively) orally for 28 days consecutively. The study revealed normal behaviour, no mortality, no significant changes in renal functions test and no drug related morphological changes in histo-pathological examination (Kanojia et al. 2013).

Rasnadi Guggulu
Clinical study

Lumbago: A clinical study evaluated efficacy of *Rasnadi Guggulu* and *katibasti* with *Dashmool oil* in 30 patients of lumbago. Ten patients of group A (*Rasnadi Guggulu*) have got 47.89% relief. Ten patients of group B (*Katibasti*) have got 47.3% relief and 10 patients of group C (*Rasnadi Guggulu*) have got 47–89% relief 10 pts. In group B (*Katibasti*) there was 47.3% relief and 10 patients of group C (*Rasnadi Guggulu* + *katibasti*) have got 84.05% relief (Sharma and Bandil 2008).

Rasnasaptaka Quatha
Clinical study

Rheumatoid arthritis: Thirty patients suffering from *Amavata* between 15–70 years of age were randomly selected for the trial. They were administered oral drug of *Rasnasaptaka Quatha* 40 ml with 3 grams of *Pippali churna* twice daily before meals for 45 days and reviewed once in 15 days. Relief in pain, improvement in appetite, increase in Hb%, RA factor, ASO titer, reduction in inflammation and endurance, etc., factors have been statistically evaluated and the results were encouraging (Anuradha et al. 2013).

Rasona Pinda

Clinical study

***Rheumatoid arthritis*:** In group I (*Rasona Pinda*), 27 patients completed the study of a total of 33 patients registered in the group (six patients dropped out mid-therapy). In group II (control group), 23 patients completed all three follow-ups out of 30 patients (there were seven dropouts in mid-therapy). In group I, complete remission in 29.6%, major improvement in 59.3% and minor improvement 11.1% were observed. In group II, complete remission in 13%, major improvement in 21.7%, minor improvement in 39.1% and unchanged condition in 26.9% of the patients was observed (Singh et al. 2010).

Rohitakarista

Pre-clinical pharmacology

***Antihyperlipidemic*:** *Rohitakarista* was administered per oral route in albino rats at a dose of 100 mg/kg body weight, once daily, up to 46 days for all the experiments. Forty rats, equally of both sexes, were randomly grouped into four where one male and one female group were used as control and other groups were used as test. In both of the male and female rats, there was a statistically very high significant decrease in the triglycerides ($p = 0.001***$). On the contrary, a statistically very highly significant increase in the total cholesterol ($p = 0.001***$), VLDL ($p = 0.001***$) and HDL ($p = 0.001***$) was noted (Obayed et al. 2010).

Clinical study

***Cholelithiasis*:** A clinical study reported efficacy of *Rohitakarista* in 20 patients of gall stones (Panigrahi et al. 2004).

Sameera Pannaga Rasa

Pre-clinical pharmacology

***Eosinophilia*:** An experimental study reported efficacy of *Sameera Pannaga Rasa* in the treatment of eosinophilia (Kansal and Dube 1980).

Clinical study

***Bronchial asthma*:** In a study, 52 patients were treated with *Sameera Pannaga Rasa* at a dose of 30 mg twice a day for four weeks along with leaf of *Piper betel* Linn. A significant improvement in subjective parameters, control on asthma, recurrence of asthma, increase in peak expiratory flow rate, considerable decrease in total and absolute, acute eosinophil count and erythrocyte sedimentation rate were observed. Overall marked improvement was found in 33.33%, moderate improvement in 44.44% and mild improvement in 20.00% was observed (Mashru et al. 2013).

Samsharkara Churna

A clinical study was carried out to study efficacy of *Samsharkara churna*. In the present study 60 patients are studied. The results were statistically analyzed with the help of chi-square test and the results have shown that the *Samsharkara churna* has good results in the *kaphaj kasa* (Dighe 2013).

Sanjivani Vati

Standardization: Quantitative estimation of tannins was done by Folin Denis method using gallic acid as standard. Embelin and Piperine were estimated by reverse phase HPLC. A standard laboratory reference sample of *Sanjivani vati* and two marketed samples were evaluated as per the developed method. Results indicated that only one marketed sample complied with all the standards prescribed and its content of tannin, piperine and embelin were equivalent to standard reference values (Parameswaran and Mandar 2010).

Clinical study

Irritable bowel syndrome: A study reported efficacy of *Sanjivani vati* in the treatment of IBS (Babbar 2014).

Saptamrita Lauha

Clinical study

Mopia: In the present study, a total 60 patients with age group between 8 to 30 years were selected randomly and were divided in two groups. In Group A, *Saptamrita Lauha* 250 mg twice daily with unequal quantity of honey and *Ghrita* were administered while in Group B, patients were subjected to Yoga therapy for three months duration with 1 month follow-up. There was no significant reduction in the visual acuity and clinical refraction, but associated changes were observed as reduced in group B when compared to group A (Bansal 2014).

Saraswataristam

Standardization: The physico-chemical studies showed total ash content as 1.1%, extractive values and some trace elements such as Lead, Mercury, Cadmium and Arsenic with 3.1, 0.047, 0.17 and 0.46 ppm respectively and all values are found within the acceptable limits specified by WHO. FTIR and HPTLC studies showed the presence of asiaticoside in *Saraswataristam*, resulting in its chemical standardization.

Preclinical pharmacology: *Saraswataristam* showed signs of dose dependent significant (P < 0.001) reduction in various episodes of epileptic seizures in comparison with standard phenytoin, thereby making it biologically standardized.

Sarva-Juara-Hara Lauha

Clinical study

Iron deficiency anemia: The results showed that there was statistically significant rise (p < 0.001) in all of them – Hb, PCV, TRBC, MCV, MCH MCHC and plasma iron, percent saturation and plasma ferritin. Total iron binding capacity decreased significantly (p < 0.001). The Hb regeneration rate was 0.16 g/dl/day for *Sarva-Juara-Hara Lauha* (Sharma et al. 2007).

Shadbindu Taila

Standardization: Chemical characterization of *Shadbindu Taila* was done on the basis of kaempferol using validated-High Performance Thin Layer Chromatographic method. *Shadbindu Taila* did not show presence of any of the heavy metals analyzed and was found non-irritant on rabbit skin (Shailajan et al. 2013).

Shankha Bhasma (Conch shell calx or Calcified conch shell)

Preclinical pharmacology

Gastroprotective: *Shankha bhasma* caused significant reduction in ulcer index (P < 0.001) in both the indomethacin and cold restraint models. TBARS of stomach in indomethacin treated rat were also reduced (P < 0.001) by *Shankha bhasma* but the serum calcium level was not altered (Pandit et al. 2000).

Clinical study

Hyperacidity: Efficacy of *Shankha bhasma* and *Amalaki churna* in the management of hyperacidity (Tank et al. 2005).

Shankhapani rasa

Clinical study

Dysmenorrhoea: Thirty patients were given *Shankhapani Rasa* in a dose of 125 mg capsules twice in a day for one month randomly and the effect was evaluated on pre-test and post-test design. Statistically significant (p < 0.01) results were seen in subjective symptoms like pain during menstruation and duration of pain (Satya Priya 2013).

Shivagutika

Clinical study

Polycystic ovarian disease: *Shivagutika* was administered twice daily with honey after food intake for 60 days. In most of the cases, there was a progressive reduction in the symptoms and size of the cyst as evident in the ultrasonography with time, indicating the efficacy of the formulation in cystic ovaries (Vishwesh and Dwivedi 2014).

Satavari Mandur

Preclinical pharmacology

Antiulcerogenic: The ulcer protective effect of *Satavari Mandur*, 125–500 mg/kg given orally, twice daily for three, five and seven days, was studied on cold restraint stress-induced gastric ulcer in rats. The effective regimen was found to be 250 mg/kg given for five days and hence was used for further experiments. *Satavari Mandur* showed significant protection against acute gastric ulcers induced by pyloric ligation but was ineffective against aspirin- and ethanol-induced ulcers (Datta et al. 2002).

Clinical study

Peptic ulcer: A clinical trial was conducted on patients having uncomplicated peptic ulcer (nine duodenal ulcer cases and one case of linear ulcer at lower end of esophagus). *Satavari Mandur* was advised to be taken in a dose of 1.5 g two times a day along with the meal. At the end of the trial, various symptoms of peptic ulcer were relieved (Vani and Tiwari 2002).

Shatapushpadya Churna

Standardization: The presence of bottle neck shaped stone cells, Aleurone grains of Maricha, Stratified fibres, Oil globules, Prysmatic crystals of Shatapushpa, Scleroids, Stone cells of Vidanga were the characteristic features observed in the microscopy of drug combination. Pharmaceutical analysis showed that Loss on drying 2.95% w/w, Ash value 25.58% w/w, Acid insoluble ash 2.9% w/w, pH 5.5 (Deshmukh et al. 2014).

Sheetamshu Rasa

Preclinical pharmacology

Antimicrobial: The antimicrobial activity of *Sheetamshu rasa* was tested against gram-positive and gram negative organisms and fungi using Kirby-Bauer and Stokes' disc diffusion method. *Sheetamshu rasa* showed remarkable antibacterial activity against *E. coli*, *Pseudomonas aeruginosa*, *Staphylococcus aureus* and *Klebsiella pnemoniae* and antifungal activity against *Candida albicans* (Raghuveer and Rao 2014).

Shila Sindoor

Toxicity: In this study, 12 rats were selected randomly from a stock colony and divided into two equal groups of six rats each in group-I (vehicle control) and group-II (treatment group). The results suggested that body weight, 13 different blood parameters like WBC, RBC, platelet count, etc., 8 types of biochemical parameters like SGOT, SGPT, creatinine, urea, etc., and lipid-peroxidation of group-II were not statistically significant in comparison with group-I. The histopathological study of kidney and liver of both groups revealed normal histology (Srilakshmi et al. 2013).

Shringa Bhasam (Deerhorn calx)

Standardization: Preliminary work on standardization of *Shringa Bhasam* has been reported (Rajasekaran and Murugesan 2002).

Shilajit (Mineral pitch or mineral wax)

Occurrence: *Shilajit* is found in the Tibet, Altai and Caucasus Mountains, as well as the Gilgit Baltistan region of northern Pakistan, but it is found most commonly in the Himalaya Mountains. Discovery of *Gomutra Shilajit* has been reported from South India (Hemadri 1987).

Mineral origin: *Shilajit* is obtained as exudate from rocks (Rajagopalan 1984; Ghosal 1992). The exudate contains 50% pure *Shilajit* and the rest are impurities. Sometimes *Shilajit* is obtained in pure form (Rajnath and Prasad 1942). Shilajit has been described as bitumen varying greatly in consistency from a free flowing liquid to hard brittle solid; a mineral resin, a plant fossil exposed by the elevation of the Himalayas, a substance of mixed plant and animal origin and an inorganic material (Tewari et al. 1973).

Vegetable sources of Shilajit

***Euphorbia royleana* Boiss.** (Euphorbiaceae): *Shilajit* has been regarded as the latex of a species of Euphorbia (Rajnath and Prasad 1942; Lal et al. 1988). *Shilajit* is reported to contain a large number of organic compounds; the analysis of the latex of *Euphorbia royleana* Boiss., collected in the summer months from the plant growing in the vicinity of *Shilajit* exuding rocks of the U.P. Himalayas, revealed the presence of identical organic compounds; this strong evidence proves that the chemical constituents of *Shilajit* are primarily derived from *E. royleana* (Ghosal et al. 1976).

***Styrax officinalis* L.** (Styraceae): *Shilajit* is thought to be largely the result of humification of the plant *S. officinalis*.

***Trifolium repens* L.** (Fabaceae): *T. repens* has also been found growing abundantly in the vicinity of *Shilajit*-bearing rocks and are responsible, at least in part, for the formation of *Shilajit* (Ghosal 1988b).

Bryophytes: Research claims that the mosses of species (Barbula, Fissidenc, Minium, and Thuidium) and species of Liverworts (Asterella, Dumortiera, Marchantia, Pellia, Plagiochasma, Stephenrencella and Anthoceros) were present in the vicinity of *Shilajit*-exuding rocks and these bryophytes are responsible for the formation of *Shilajit* (Joshi et al. 1994).

Chemistry

Shilajit is composed mainly of humic substances, including fulvic acid (60% to 80%), humic acid, benzoic acid and hippuric acid, elements including selenium, fatty acids, resins, latex, gums, albumins, triterpenes, 3,4-benzocoumarins, sterols, aromatic carboxylic acids, 3,4-benzocoumarins, amino acids, polyphenols, and phenolic lipids (Ghosal et al. 1976; Kong et al. 1987; Ghosal et al. 1991; Ghosal 1994; Ghosal 1995; Ghosal, Kawanishi and Saiki 1995; Ghosal, Lata and Kumar 1995). The Government Laboratory Sydney (2003) has found 67 trace minerals in significant quantities.

Structure of Fulvic acid

Work of Ghosal (1990) on chemical study of *Shilajit* is remarkable. The constituents of *Shilajit* are of two types:

1. Humic: They constitute 80–85% mass of *Shilajit*. These are produced by interaction of *Shilajit* with lower plants (algae, mosses, and liverworts), micro-organisms and even higher plants.
2. Non-humic: They constitute 8–10% mass of *Shilajit*. Non-humic substace s includes low molecular weight chemical compounds (acuparins), oxygenated dibenzo-α-pyrones and triterpene acids of the tirucullane type.

Recently 4á, 5á,6á-trihydroxygeranyl acetat; 6-(9,9-dimethylbutyl) phenol; 1-cyclohexyl-3, 4-dihydroxybenzene; 2, 3, 12, 13-tetrahydroxy-10, 15[a, f] – phenyl xanth-17-one; 2, 3, 13, 14-tetrahydroxy-15, 16-[a, f] – phynyl-7H-anthracen-18-one and 3-hydroxynaphthalenyl-6, 7-ã-lactone have been isolated (Indu et al. 2006).

Mishra in his book on *Rasa Shastra* has given excellent comparison of chemical composition of impure and pure *Shilajit*.

S. No.	Contents	Impure	Pure
1.	Moisture	12.54%	27.03%
2.	Benzoic acid	6.42%	8.58%
3.	Hippuric acid	5.53%	6.13%
4.	Fatty acids	2.01%	1.36%
5.	Resin and wax	3.28%	2.48%
6.	Gum	15.57%	17.32%
7.	Albuminoids	17.61%	16.12%
8.	Foreign matter	28.52%	2.15%

Three samples of *Shilajit* collected from Ladak were subjected to comparative chemical studies. Presence of urea and glycine was noted in all three samples (Alam et al. 1983).

Standardisation: The sample extracts of *Shilajit* were prepared and absorbance was taken at four different wavelengths 280 nm, 260 nm, 472 nm and 664 nm. From the obtained absorbance results, ratio was calculated, where Q1 260/664, Q2 472/664, Q3 280/472 shows strongly humified material, humification index and proportion between Lignin and other material respectively. The humification index indicates the presence of fulvic acid (Renuka et al. 2013).

Pharmacology

Anti-anxiety: *Shilajit* at a dose of 10 mg/kg, p.o. has significant anti-anxiety activity as proved by elevated plus-maze test which is comparable to that of diazepam (1 mg/kg, p.o.) (Jaiswal and Bhattacharya 1992).

Antidiabetic: A study reported effect of *Shilajit* on glucose tolerance in rats (Gupta 1966). *Shilajit* (50 & 100 mg/kg, p.o.) had no discernible per se effect on blood glucose levels in normal rats but attenuated the hyperglycemic response of streptozotocin (Bhattacharya 1995a).

Fulvic acids showed significant success in preventing and combating free radical damage to pancreatic islet B cells, which is the widely accepted cause for diabetes mellitus (Bhattacharya 1995b). In yet another study, *Shilajit* potentiated the hypoglycaemic action of insulin and inhibition of streptozotocin induced diabetes in rats (Kanikkannan et al. 1995).

Anti-nociceptive: *Shilajit* has inhibitory effects on the substantia gelatinosa neurons of trigeminal subnucleus caudalis (Vc) through chloride ion channels by activation of the glycine receptor and GABA (A) receptor, indicating that *Shilajit* contains sedating ingredients for the central nervous system (Yin et al. 2011).

Anti-fatigue: Anti-fatigue effect of processed *Shilajit* standardized to dibenzo-a-pyrones (0.43% w/w), DBP-chromoproteins (20.45% w/w) and fulvic acids (56.75%

w/w) was in a rat model. The results indicate that *Shilajit* mitigates the effects of chronic fatigue syndrome in this model possibly through the modulation of hypothalamus–pituitary–adrenal (HPA) axis and preservation of mitochondrial function and integrity (Surapanenia et al. 2012).

Anti-inflammatory: Orally administered *Shilajit* (50 mg/kg) induced significant anti-inflammatory activity against carrageenan induced pedal oedema (Acharya et al. 1988). The alcoholic and aqueous extract of *Shilajit* exhibited anti-inflammatory activity in carrageenam induced paw volume using indomethacin as standard (Indu et al. 2006).

Antifungal: Methanolic extract of *Shilajit* at the concentration of 5000 µg/ml was having excellent inhibitory activity against *Alternaria cajani* (95.12% spore inhibition) (Shalini et al. 2009).

Anti-lipid-peroxidative: The effects of *Shilajit* on lipid liver homogenate were investigated. It inhibited lipid peroxidation induced by cumene hydroperoxide and PDP/Fe++ complex in a dose dependent reduced glutathione content and inhibited ongoing lipid peroxidation, induced by these agents immediately after addition to the incubation system (Tripathi et al. 1996).

Antimicrobial: U.V. treatment *Shilajit* showed the potentially good antimicrobial activity against all available bacterial strains and antifungal activity only against Penicillium chrysogenum. MIC was 2 mg/ml for *Bacillus subtilis*, *Bacillus cereus*, *Escherichia coli*, *Proteus vulgaris*, 5 mg/ml for *Pseudomonas aeruginosa* and *Klebsiella pneumonia* and 3 mg/ml for *Staphylococcus aureus* (Shadab et al. 2014).

Antioxidant: *Shilajit* showed free radical scavenging & antioxidant effect against SO3-, OH radical and paramagnetic nitric oxide depending on its concentration.

Analgesic: Studies were conducted in albino mice to determine the effect of 50–200 per kg of *Shilajit*. The analgesic effect of *Shilajit* pretreatment were studies using the technique of hot wire induced tail-flick response. *Shilajit* was found to have analgesic activity (p, 0.001) in the dose of 200 mg/kg i.p. The effect was significant during the first 60 min (Acharya et al. 1988).

Antiulcerogenic and ***anti-inflammatory***: *Shilajit* increased the carbohydrate/protein ratio and decreased gastric ulcer index, indicating an increased mucus barrier. *Shilajit* was found to have significant anti-inflammatory effect in carrageenan-induced acute pedal oedema, granuloma pouch and adjuvant-induced arthritis in rats (Goel et al. 1990). Antiulcerogenic activity of fulvic acids and 4"-methoxy-6-carbomethoxybipheny isolated from *Shilajit* has been reported (Ghosal et al. 1988).

Cardioprotective: An investigation explored effect of mumie (*Shilajit*) pre-treatment on cardiac performance of rats subjected to myocardial injury. *Shilajit* was given at dosages of 250 and 500 mg/kg/day, orally for seven days, respectively to animals divided into control, M250, and M500. Isoproterenol (85 mg/kg i.p.) was injected (s.c.) on the 6th and 7th days, to half of the animal subgroups to induce myocardial damage.

Mumie pre-treatment had no significant effects on hemodynamic and cardiac indices of normal animals. When the cardiac injury was induced, mumie maintained the ±dp/dt maximum, attenuated the serum cardiac troponin I, and reduced the severity of cardiac lesions. Antioxidant system seems to be a probable mechanism (Joukar et al. 2014).

***Immunomodulator*:** It was found that the white blood cell activity was increased by *shilajit* extract. The observed activity increased as the dose of *Shilajit* extract and time of exposure was increased (Bhaumik et al. 1993).

***Mast cell stabilizing*:** *Shilajit* and different combination of its constituents provided statistically significant protection to antigen-induced degranulation of sensitized mast cells, markedly invited the antigen-induced spasmin of sensitized guinea pig ileum and prevent mast cell disruption induced by compound 48/80 (Ghosal et al. 1989).

***Morphine-tolerance*:** Chronic administration of morphine (10 mg/kg, i.p., b.i.d.) to mice over a duration of 10 days resulted in the development of tolerance to the analgesic effect of morphine. Concomitant administration of PS with morphine, from day 6 to day 10, resulted in a significant inhibition of the development of tolerance to morphine (10 mg/kg, i.p.) induced analgesia (Tiwari et al. 2001).

***Neuroprotective*:** Diffuse traumatic brain trauma was induced in rats by drop of a 250 g weight from a 2 m high (Marmarou's methods). Animals were randomly divided into five groups including sham, TBI, TBI-vehicle, TBI-Shi150 group and TBI-Shi250 group. Rats had undergone intraperitoneal injection of *Shilajit* and vehicle at 1, 24, 48 and 72 hr after trauma. Brain water and Evans blue dye contents showed significant decrease in *Shilajit*-treated groups compared to the TBI-vehicle and TBI groups. Intracranial pressure at 24, 48 and 72 hr after trauma had significant reduction in *Shilajit*-treated groups as compared to TBI-vehicle and TBI groups ($P < 0.001$) (Khaksari et al. 2013).

***Nootropic*:** *Shilajit* at a dose of 50 mg/kg, p.o. has significant nootropic activity as shown by passive avoidance learning and retention (Jaiswal and Bhattacharya 1992). *Shilajit* and extract from *Withania somnifera* affect preferentially events in the cortical and basal forebrain cholinergic signal transduction cascade. The drug-induced increase in cortical muscarinic acetylcholine receptor capacity might partly explain the cognition-enhancing and memory-improving effects of extracts from *W. somnifera* observed in animals and humans (Schliebs et al. 1992).

***Spermatogenic* and *ovogenic*:** *Shilajit* was administered orally to 7-week-old rats over a 6-week period. In the male rats, the number of sperms in the testes and epididymides was significant higher than in the control. A histological examination revealed an apparent increase in the number of seminiferous tubular cell layers in the testes of the treated rats. In the female rats, the effect of *Shilajit* was estimated by the ovulation inducing activity. Over a 5-day period, ovulation was induced in seven out of nine rats in the *Shilajit* administration group and in three out of nine rats in the control (Park et al. 2006).

Clinical study

Obesity: In the present study, 66 patients of obesity were treated with *Shilajatu* processed with *Agnimantha*. After compilition of therapy, 5.09 ± 0.24 kg and 2.06 ± 0.10 kg/m (2) reduction of body weight and body mass index respectively were noted. The result was found to be statistically highly significant (P < 0.001) (Pattonder et al. 2011).

Oligozoospermia: Initially, 60 infertile male patients were assessed and those having total sperm counts below 20 million ml (–1) semen were considered oligospermic and enrolled in the study (n = 35). PS capsule (100 mg) was administered twice daily after major meals for 90 days. Twenty-eight patients who completed the treatment showed significant (P < 0.001) improvement in spermia (+37.6%), total sperm count (+61.4%), motility (12.4–17.4% after different time intervals), normal sperm count (+18.9%) with concomitant decrease in pus and epithelial cell count compared with baseline value. Significant decrease of semen mlondialdehyde content (–18.7%) was observed.

Toxicity: Administration of *Shilajit* in different dose levels in rats for 91 days revealed no significant changes in iron level of treated groups when compared with control except liver (5,000 mg/kg) and histological slides of all organs. The weight of all organs was normal when compared with control (Velmurugan et al. 2012).

Administration of two gms of *Shilajit* for 45 days did not produced any significant change in physical parameters, i.e., blood pressure, pulse rate and body weight and similarly no charge was observed in hematological parameters. A signification reduction in Serum Triglycerides, Serum cholesterol with simultaneous improvement in HDL Cholesterol was seen, besides which *Shilajit* also improved antioxidant status of volunteers (Sharma et al. 2003).

Shirishavaleha

Pre-clinical pharmacology

Anti-inflammatory: *Shirishavaleha* was administered orally at a dose of 1.8 g/kg for five days. Phenylbutazone was used as the standard anti-inflammatory drug for comparison. Between the two different test samples studied, the formulation made from heartwood showed weak anti-inflammatory activity in this model while that made from the bark produced a considerable suppression of edema after 6 h (Yadav et al. 2010).

Immunomodulatory: *Shirishavaleha* prepared from heartwood shows significant enhancement in antibody formation, attenuation of body weight changes, and suppression of immunological paw edema, while *Shirishavaleha* prepared from bark shows weak immunomodulatory activity (Yadav et al. 2011).

Shivaksharpachan Churna

Pre-clinical pharmacology

Gastroprotective: Ethanolic extract of *Shivaksharpachan Churna* (50, 100 and 200 mg/kg body weight) were administered orally, twice daily for five days for prevention from pylorus ligation and ethanol-induced ulcers followed by estimation of anti-oxidant enzymes. Ethanolic extract of *Shivaksharpachan Churna* prevents the oxidative damage of gastric mucosa by blocking lipid peroxidation and significant increase in superoxide dismutase and catalase activity (Patra et al. 2011).

Sidh Makardhwaj and *Makardhwaj*

Toxicity: *Makardhwaj* was administered to a group of experimental rabbits in the dose of 10 mg per kg body weight daily for one year. The decrease in plasma and R.B.C. volume was found to be less in the Makaradhvaja treated group than in the control series. While creatinine clearance was reduced in the control series, no change was observed in the Makaradhvaja treated group (Shukla et al. 1992). *Makardhwaj* was without any toxic potential even at the dose of 2000 mg/kg in animals equivalent to 22.4 g for human being (Ravishankar and Ravi 2012).

Graded doses of *Sidh Makardhwaj* (10, 50, 100 mg/kg) did not cause significant change in neurobehavioural parameters, brain cerebrum AChE activity, liver and kidney function tests as compared to control. The levels of mercury in brain cerebrum, liver, and kidney were found to be raised in a dose dependent manner (Kumar et al. 2014).

Simhanada Guggulu

Clinical study

Rheumatoid arthritis: In the present study, 24 patients of *Amavata* were registered and randomly grouped into two. In group A, *Shiva Guggulu* 6 g/day in divided doses and in group B, *Simhanada Guggulu* 6 g/day in divided doses were given for eight weeks. On analysis of the results, it was found that *Simhanada Guggulu* provided better results as compared to *Shiva Guggulu* in the management of *Amavata* (Pandey et al. 2012).

Total 101 patients of *Amavata* were registered for the present study and were randomly divided into two groups. In group A-*Rasona Rasnadi Ghanavati* 2 Vati thrice/day was given for three months, while in group B-*Simhanada Guggulu* 2 Vati thrice a day for three months was administered. Along with this, *Rasona Rasnadi Lepa* was applied locally over affected joints twice daily in both groups. The results of the study showed that both the groups showed significant relief in symptoms; however, compared to *Simhanada Guggulu*, *Rasona Rasnadi Ghanavati* showed better result in the management of *Amavata* (Mahto et al. 2011).

Sitopaladi Churna

Standardization: Four marketed preparations and in-house preparations of *Sitopaladi Churna* were used for the study. The various parameters including organoleptic characteristics and HPTLC was carried out for quantitative analysis of all the formulations (Panda et al. 2012).

Preclinical pharmacology

Anti-allergic: A study investigated the effect of *Sitopaladi Churna* on IgE mediated allergic reactions both *in vivo* and *in vitro*. Results of studies employing RBL-2H3 cell line and rodents respectively, revealed that SC-1 inhibits IgE mediated reactions significantly (Rayees et al. 2012).

Anti-inflammatory: *Sitopaladi churna* extract exhibited good anti-inflammatory effects in rats, causing a dose-related inhibition of the increase in the paw circumference (acute inflammation) induced by subplantar injection of fresh egg albumin and carrageenin. It also significantly decreased the weight of cotton pellet (chronic inflammation). It also protected mast cell disruption induced by compound 48/80 (Bharti et al. 2008).

Antitussive: The standard drug codeine phosphate brought about a reduction of bouts of cough from 15.62 ± 0.38 to 1.0 ± 0.11 (93.60% inhibition), which was significant ($P < 0.01$). The percentage inhibition of bouts of cough for in-house formulation (94.28%) was very significant compared to the standard as well as other marketed formulations (Pattanayak et al. 2010).

Mast cell stabilization: Effect of *Sitopaladi Churna* on mast cells degranulation was studied by administering the churna in rats. The mast cell stabilizing activity can be attributed to a great extent due to the presence of piperine (Makhija et al. 2013).

Srngyadi Churna

Standardization: *Srngyadi Churna* was prepared in an institute pharmacy as per Ayurvedic formulary of India, Part-I guide lines and attempts to evaluate the organoleptic characters, pharmacognostic study and physicochemical parameters like pH, Loss on drying at 105°C, Water soluble extract, Alcohol soluble extract, Total Ash, Acid insoluble ash and TLC (Meena et al. 2013).

Sveta parpati

Clinical study

Urolithiasis: A study reported efficacy of *Sveta parpati* with *Pasana bheda* and *Gokshura* in the management of urolithiasis (Sannd et al. 1993).

Sudarshan Churna

Preclinical pharmacology

Antimicrobial: Aqueous extract of *Sudarshan Churna* was found active against the gram-negative bacteria K. pneumoniae, *E. coli*, and gram-positive bacteria *S. aureus*, *P. vulgris* and found less effective against gram-positive bacteria S. epidermidis and B. subtilis. Aqueous extract of *Sudarshan Churna* was shown as significantly less effect against *C. Albicans* (Bhargava 2008).

Sutshekhar Rasa

Preclinical pharmacology

Gastroprotective: In cold restraint induced ulcers model, *Kamadugha* and *Sutshekhar Rasa* significantly inhibited the formation of gastric lesions induced by cold restraint-stress by 74.37% and 74.83% respectively. Animals in the pylorus-ligation group showed a significant increase in the ulcer index and acid secretory parameters like gastric volume, pH, free and total acidity when compared with those of vehicle treated group. *Kamadugha* and *Sutshekhar Rasa* at a dose of 40 mg/kg body weight each showed protection index of 47.59% and 50.69%, respectively in ethanol-induced ulcers model (Chandra et al. 2010).

Swarna Basant Malti

Chemical investigations: In the present trial the presence of gold have been verified in the medicine through tests (Dube and Sharma 1980).

Clinical study

Pulmonary tuberculosis: In the clinical trial of *Swarna Basant Malti* in dose of 170 mg t.d.s. with honey, the effect was observed on patients of pulmonary tuberculosis and was found most encouraging when combined with first line drugs. The most appreciable effect which was observed was in regimen II which suggests that the drug may be a good adjuvant. No acute or chronic toxicity of the drug was noticed in the follow up of the cases up to one year. Resistance to thiacetazone was minimum in fresh cases and 3% in post treated cases. Resistance to I.N.H. was higher and less than what was found to streptomycin.

Swarna Bhasma (Incinerated Gold)

Chemical Analysis: Qualitative analyses indicated *Swarna bhasma* contains gold and microelements including Fe, Al, Cu, Zn, Co, Mg, Ca, As, Pb, etc. (Mitra et al. 2002).

Preclinical pharmacology

Analgesic: *Swarna Bhasma* and *Kushta Tila Kalan* (25–50 mg/kg, p.o.) and Auranofin (2.5–5.0 mg/kg, p.o.) exhibited analgesic activity against chemical (acetic acid induced writhing), electrical (pododolorimeter), thermal (Eddy's hot plate and analgesiometer) and mechanical (tail clip) test. While the analgesic effects of *Swarna Bhasma* and *Kushta Tila Kalan* could be partly blocked by pretreatment with naloxone (1–5 mg/kg, i.p., –15 min), such antagonism was not discernible with Auranofin at the doses used (Bajaj and Vohora 1998).

Anti-depressant, anti-anxiety and *anti cataleptic*: The test drugs caused significant increase in punished drinking episodes in anxiometer and open arm entries and time in elevated plus maze and decrease in behavioural deficit. A decrease in immobility time in forced swimming test, normalization of shock induced escape failures in learned helplessness test, and reduction of haloperidol-induced catalepsy scores were also noted in treated animals. The maximum tolerated doses were found to be more than 80 times the effective doses and no weight loss or untoward effects were observed on gross behaviour and hematological parameters (Bajaj and Vohora 2000).

Antioxidant: In an experimental animal model, chronic *Swarna bhasma*-treated animals showed significantly increased superoxide dismutase and catalase activity, two enzymes that reduce free radical concentrations in the body (Mitra et al. 2002).

Blood compatibility studies: The *Swarna bhasma* preparations with a crystallite size of 28–35 nm did not induce any blood cell aggregation or protein adsorption. Activation potential of these preparations towards complement system or platelets was negligible. These particles were also non-cytotoxic (Paul and Sharma 2011).

Immunomodulator: *Swarna Bhasma* and *Kushta Tila Kalan* significantly ($P < 0.001$) increased counts of peritoneal macrophages and stimulated phagocytic index of macrophages. Auranofin elicited a suppressive action on these parameters.

Swarna Makshika Bhasma (Incinerated Chalcopyrite or Chalcopyrite Calyx)

Physicochemical characterization: *Swarna makshika bhasma* was analyzed using X-ray Diffraction analysis of raw bhasma. The analysis revealed that raw *Swarna makshika bhasma* contains $CuFeS_2$, and SM bhasma contains Fe_2O_3, FeS_2, CuS and SiO_2. Scanning Electron Microscope studies showed that the grains in *Swarna makshika bhasma* were uniformly arranged in agglomerates of size 1–2 microns as compared to the raw *Swarna makshika bhasma* which showed a scattered arrangement of grains of size 6–8 microns (Mohaptra and Jha 2010).

Effect on bio-chemical parameters: Rats were divided into two groups. *Swarna makshika bhasma* mixed with diluted honey was administered orally in a therapeutic dose to Group *Swarna makshika bhasma* and diluted honey only was administered to vehicle control Group, for 30 days. The blood samples were collected twice, after 15 days and after 30 days of drug administration. It was observed that Hb% was found significantly increased and LDL and VLDL were found significantly decreased in Group SMB when compared with vehicle control group (Mohaptra and Jha 2011).

Subchronic genotoxicity: The results of the study revealed a lack of generation of structural deformity in above parameters by tested drugs compared to the cyclophosphamide treated group. Observed data indicate that the *Bhasmas* tested were non-genotoxic under the experimental conditions (Savalgi et al. 2012).

Swarna Parpati

Standardisation: A study reported method of standardisation of *Swarna Parpati* (Pandey 1982).

Swarna Vanga

Standardization: *Swarna vanga* requires mild (<250 C) and moderate fire (250–500 C) for a period of 9 h each for preparation. *Swarna vanga* with 42.9% yield was observed and having 63.2 and 34.4% tin and sulfur, respectively (Gokarn et al. 2013).

Pre-clinical pharmacology

***Hypoglycemic*:** The effect of *Swarna Vanga* with half the amount of mercury was tested in adult alloxan-induced diabetic albino rats; the effects were not statistically significant (Suresh et al. 1988).

***Testicular regeneration*:** The ability of *Swarna vanga* in generation of testicular germinal epithelium is seen on partial damaged testis which is produced by inducing CdCl2 (Singh et al. 2012).

Toxicity: *Swarna vanga* does not show any toxic effects in the therapeutic dose (12.5–25 gm/body weight of albino rats) but shows some toxic effects in higher dose in longer duration (Singh et al. 2012). An animal experiment was carried out for short duration (14 days) to screen the toxic effects of *Swarna vanga* in increasing doses of the drug starting from the maximum therapeutic dose (12.5 mg/100 gm b.wt./day). The drug was found to have no toxic effects in tissues of the animal at doses of 12.5 mg and 25 mg/100 gm b.wt./day observed (Sharma et al. 1985).

Swas Kasa Chintamani Rasa

Clinical study

***Childhood Bronchial Asthma*:** A total of 23 children (who had consented) of both sex under 12 years of age were included in the study and divided into three groups, blood samples were collected before treatment and after the completion of therapy for the metabolic markers like Hb gm%, TLC, AEC, S. Protein, S. Albumin, SGOT, SGPT, alkaline phosphatase and S. Bilirubin. SKCR was given for a total of 45 days in a dose of 4 mg/kg/dose × 12 hourly with garlic, ginger and honey in ratio of 1:2:4. Statistical

Analysis Used: In the present study, SPSS software was used to get statistical data such as Mean (X-), Mean Difference (d'), Standard Deviation (SD) and Student's "t" test, etc. (Kumar, Singh and Gupta 2014).

Swasakuthar Rasa

Pre-clinical

Antibacterial: *Swasakuthar rasa* was effective against only three strains of *S. aureus* and one strain did not show any effect of it. *Swasakuthar rasa* shows different minimum inhibitory concentration on *S. typhimerum, Pseudomonas, Morganella, Shingella, Serratia, S. aureus* and *E. coli* was 4.5 mg/ml, 75 mg/ml, 150 mg/ml, 18.5 mg/ml, 300 mg/ml and 9.25 mg/ml respectively (Das et al. 2012).

Surya Shekhara Rasa

Standardization: Values of Physico-chemical studies of all three samples of *Surya Shekhara Ras* were almost in the same range HgS compound form was detected in X-RD study of three samples with Hexagonal crystal structure. Inductive Coupled Plasma with Optical Emission Spectrometry showed presence of 94% Hg. FT-IR analysis reveals the presence of similar organic functional groups in three samples of *Surya Shekhara Ras* (Arya et al. 2013).

Sunder Vati

Clinical study

Acne vulgaris: Eighty two patients with acne vulgaris were randomised into five groups. Four different Ayurvedic treatment schedules were given orally for six weeks, while one group received a placebo. Physical and clinical investigations were carried out at two week intervals. A significant reduction in lesion count was observed in patients receiving *Sunder Vati* when compared with the placebo and the other Ayurvedic formulations, which failed to produce any significant difference from the pretreatment condition. The drug therapies were well tolerated (Paranjpe and Kulkarni 1995).

Talisadya Churna

Analytical study: Thin layer chromatographic patterns have been evolved for identifying the presence of the ingredients along with the general parameters like loss on drying, ash value, water and alcohol soluble extractives, volatile oil content, sugar and silica content (Dave and De 1996).

Standardization: *Talisadya Churna* prepared by grinding the ingredients in mixie showed less acid insoluble content, high volatile matter, water soluble matter, and exhaustive extraction in chloroform. Thin layer silica gel chromatography and test of organic functional groups did not show any difference in the *Talisadya Churna* prepared by either method (Alam et al. 1991).

Tamra Bhasam (Incinerated Copper or Copper calyx)

Chemical Evaluation: It was seen that samples from different batches and different manufacturers contained varying amounts of *Tamra Bhasma*. Spectrophotometric analysis showed that copper was present predominantly in its cupric state. X-Ray Diffraction analysis showed that the copper in the bhasma was present in its sulphide form (Chitnis and Stanley 2011).

Preclinical pharmacology

Anti-hyperlipidemic: The hyperlipidemia was induced by feeding high fat diet in Wistar strain albino rats. The parameters including body weight, weight of various organs, serum lipid profile and histopathology of liver, kidney, heart and aorta were studied. *Tamra Bhasma* prepared from purified Tamra which has significant anti-hyperlipidemic activity, while unpurified *Tamra Bhasma* is devoid of such effects (Jagtap et al. 2013).

Antioxidant: *Tamra Bhasma* was orally given for 7, 15 and 30 days in different doses. The best protective response was found at the dose of 0.5 mg/100 g body weight in albino rats, although it showed some histopathological changes at the dose of 20 mg/100 g body weight (Pattanaik et al. 2003).

Hepatoprotective: *Tamra Bhasma* was orally given for eight days which showed significant reduction in the level of malondialdehyde production at different concentrations of cumene hydroperoxide *in vitro*. Glutathione content was maintained upto seventy minutes and SOD activity was enhanced to 166%. These animals did not show any rise in serum GOT and GPT. On similar doses no histological changes were observed in liver (Tripathi and Singh 1996).

Gastroprotective: In the present study the anti-ulcerogenic effect of *tamra bhasma* was observed in 8-h immobilised, 4-h pylorus-ligated, and aspirin-induced gastric ulcers in rats. The anti-ulcerogenic effect of the drug was also studied in histamine-induced gastric and duodenal ulcers in male guinea pigs. The minimal oral effective anti-ulcerogenic dose of *Tamra Bhasma* has been determined to be 1 mg/kg. The drug in this dose caused a decrease in the total acid and pepsin output and an increase in the carbohydrate/protein ratio, indicating increased mucus secretion in the gastric secretion of rats (Sanyal et al. 1983).

Effect on ponderal and biochemical parameters: *Tamra Bhasma* and *Somnathi Tamra Bhasma* was administered to wistar strain albino rats for 45 consecutive days. Blood was collected and rats were sacrificed on the 46th day. Results showed significant decrease in serum cholesterol, high density lpoprotein cholesterol, triglycerides, total protein, and serum alkaline phosphatase levels (Chaudhari et al. 2014).

Toxicity: To assess the toxicity of *Tamra Bhasma* prepared with various methods, different samples of *Tamra Bhasma* were prepared by following different procedures, all the four prepared samples were subjected for acute & sub acute toxicity study by following Staircase method on Albino rats. The results significantly suggest that *Somanathi Tamra Bhasma* was less toxic and few samples of *Tamra Bhasma* showed mild to moderate toxic signs & symptoms which are reversible (Honwad 2011).

In another study, *Tamra bhasma* prepared from purified and unpurified *Tamra* was subjected to an oral toxicity study to ascertain the role of *Shodhana* process on safety profile of *Tamra bhasma* on subchronic administration to albino rats. *Tamra bhasma* prepared from unpurified Tamra has pathological implications on different hematological, serum biochemical and cytoarchitecture of different organs even at the therapeutic dose level (5.5 mg/kg). Whereas, *Tamra bhasma* prepared from *Shodhita Tamra* is safe even at five-fold to therapeutic equivalent doses, i.e., 27.5 mg/kg (Jagtap et al. 2013).

Chronic toxicity was conducted in albino rats (wistar strain). In this study, *Tamra Bhasma* was administered orally daily to different groups of albino rats in TD (*Tamra Bhasma*) and 2 TD (*Tamra Bhasma* 2 x Therapeutic Doses) doses for three months. *Tamra Bhasma* was found to be relatively safe at these dose levels. There was no mortality. No significant behavioural changes were noted in any of the group studied. No major alterations were observed in haematology, serum biochemical, necropsy and histopathology at the administered dose level (Vahalia et al. 2011).

Genotoxicity: *Tamra Bhasma* did not induce micronuclei formation or an increase in the percentage of DNA damage, showing its safety upon consumption (Sathya et al. 2008).

Tankan Bhasam

Preclinical pharmacology

Chronic tonsillitis: *Tankana Bhasma* is used as a treatment in chronic tonsillitis in the form of a gargle. To ensure scientific validity of the efficacy, a comparative study was conducted between Aspirin tablet and *Tankana bhasma* which was statistically analyzed. In the study, *Tankana bhasma* showed significant relief of symptoms that were in chronic tonsillitis (Ravishankar and Mahesh 2013).

Tarakeswara Rasa

Preclinical pharmacology

Antidiabetic: Administration of *tarakeswara rasa* to alloxan induced diabetic rats resulted in a significant decrease in the blood glucose. Its antidiabetic effect was also evident from the tissue parameters estimated. The drug administration to diabetic rats also resulted in the reduction of cholesterol, triglyceride and phospholipid content (Vasanthakumari and Shyamala 1997).

Tribhuvan Kirti Rasa

Standardisation: A method has been developed to standardise *Tribhuvan kirti rasa* (Agrawal et al. 2006).

Clinical study

***Chronic sinusitis*:** *Tribhuvan kirti rasa* was administered at a dose of 250 mg b.d. with Juice of Ginger in thirty patients suffering from uncomplicated chronic sinusitis. Steam inhalation of *Dasmula kwath* was given two times a day followed by nasya of *Anu taila* at a dose of four drops in both nostrils. The duration of the treatment varied from 45 days to 90 days. The overall clinical efficacy was 96.6%. This medicine along with steam inhalation followed by *Nasya* was found to be well tolerated in general and no side effects were reported.

***Influenza*:** In a comparative clinical study involving 60 patients, 31 were administered *Tribhuvan Kirti Rasa*, 28 were administered *Hinguleshwar Rasa* and to 1, both the drugs. *Hinguleshwar Rasa* and *Tribhuvan Kirti Rasa* were effective in Influenza but *Hinguleshwar Rasa* was found more effective (Pandey 1968).

Triguna Rasa sindhoora

Standardisation: *Triguna Rasa sindhoora* was prepared in the ratio of 1:3 (Hg-1 and S-3 parts) in minimum of 15 h, both in EMF and *Valuka Yantra*. Average 603°C and 630°C temperature in EMF and Classical *Valuka Yantra*, respectively, with the yield of average 78% product in EMF and 81% product in Classical *Valuka Yantra* using 24 kg of coal (Dhundi et al. 2010).

Trikatrayadi Lauha

Clinical study

***Iron deficiency anemia*:** Safety and efficacy of *Trikatrayadi Lauha* suspension was evaluated in children with iron deficiency anemia. *Trikatrayadi Lauha* was effective to increase the hemoglobin level 1.94 g/dL (8.52–10.46 g/dL, $P < 0.001$) in five weeks and 3.33 g/dL (8.52–11.85 g/dL, $P < 0.001$) in 10 weeks. No adverse effect of the trial drug was observed during the study (Kumar and Garai 2012).

Trikatu

Pharmacognosy

Trikatu is an Ayurvedic preparation containing black pepper, long pepper and ginger (Johri and Zutschi 1992).

Standardisation

The physicochemical data and HPTLC fingerprinting profile may be utilized for laying down pharmacopoeial standards for *Trikatu Curanam* (Saraswathy et al. 1998). In this work, *Trikatu Churna* was standardized by qualitatively evaluating the preliminary phytochemicals. Piperine content was determined using HPTLC (Shailajan et al. 2011).

Pre-clinical pharmacology

Effect on bioavailability of rifampicin: Co-administration of *Trikatu* does not influence the extent of bioavailability (AUC0-infinity) but reduces the rate of bioavailability (Cmax) of rifampicin. The latter effect may reduce the efficacy of rifampicin therapy (Karan, Bhargava and Garg 1999).

Effect on bioavailability of carbamazepine: In the animals treated with single dose of *Trikatu*, there was a significant decrease in Tmax of carbamazepine ($P < 0.05$). Multiple doses of Trikatu also shortened the Tmax of carbamazepine although not to statistically significant level (Karan, Bhargava and Garg 1999).

Effect on pharmacokinetic profile of isoniazid: Co administration of *Trikatu* significantly reduced the Cmax (5.48 ñ 0.75 'g/ml vs. 8.42 ñ 0.85 'g/ml; $P < 0.05$) and AUCo-(15.04 ñ 3.64 'g/ml. hr vs. 24.76 ñ 4.03 'g/ml. hr; $P < 0.05$) of isoniazid. *Trikatu* reduces the bioavailability of isoniazid in rabbits (Karan et al. 1998).

Effect on pharmacokinetic profile of indomethacin: *Trikatu* reduces bioavailability of indomethacin (Karan, Bhargava and Garg 1999).

Effect on pharmacokinetic profile of diclofenac sodium: In rabbits, it was observed that *Trikatu* significantly decreased the serum levels of diclofenac sodium. It was observed that the mean percent oedema inhibition shown by the combination of *Trikatu* and diclofenac was similar to that shown by *Trikatu* alone but significantly less than that shown by diclofenac alone (Lala et al. 2004).

Effect on pharmacokinetic profile of pefloxacin: A study revealed a decreased plasma concentration ($p > 0.05$) of pefloxacin following *Trikatu* administration during the absorption phase (10, 15, 20 min post pefloxacin administration). In contrast, the plasma concentrations of pefloxacin were significantly higher at 4, 6, 8 and 12 h (during the elimination phase) of the pefloxacin administration (Dama et al. 2008).

Effect on bioavailability of ampicillin and norfloxacin

A study reported that coadministration of *Trikatu* and its components reduces bioavailability of ampicillin and norfloxacin in rabbits (Janakiraman and Manavalan 2009).

***Antioxidant*:** *Trikatu* mega Ext possesses statistically significance DPPH free radical scavenging activity ($P < 0.001$). Sample of 100 µg/ml inhibited the production of

Superoxide anion radical by 84.3% showing strong superoxide radical scavenging activity (Jain and Mishra 2009).

Adaptogenic: The evaluation of adaptogenic potential by oral administration of *Trikatu* mega Ext (100–200 mg/kg) evoked a significant increase in the swimming time (min) and also increases in anoxic tolerance time in physical and anoxic stress models (Jain and Mishra 2011).

Analgesic and ***anti-pyretic***: *Trikatu churna* and its individual components were tested for analgesic activity *in-vivo* hot plate method using mice. *Trikatu churna* exhibited moderate analgesic activity (Reddy and Seetharam 2009).

A study indicated that *Trikatu* exert a potent anti-inflammatory effect against monosodium urate crystal-induced inflammation in rats in association with analgesic and anti-pyretic effects in the absence of gastrointestinal damage (Murunikkara and Rasool 2014).

Trikatu treatment (1000 mg/kg/body weight) reverted back all the biochemical and immunological parameters to near normal levels in arthritic rats as evidenced by the radiological and histopathological assessments (Murunikkara and Rasool 2014a).

Anthelmintic: Piperazine citrate (10 mg/ml) was included as standard reference and distilled water as control. Time taken for induction of paralysis and death was noted. The alcoholic extract of *Trikatu churna* and its ingredients exhibited potent anthelmintic activity against *Pheritima posthuma*, but the highest activity was noticed in *Trikatu churna* (Reddy and Seetharam 2009).

Trikatu churna & each of its components were extracted with water by the process of maceration. The Albendazole suspension was used as standard. The time required for the paralysis & death was noted. All the samples possess good anthelmintic activity on *Pheritima postuma*, i.e., earthworms at their highest concentrations.

Antimicrobial: *Trikatu churna* and its individual components were tested for antimicrobial activity against certain clinical bacterial and fungal isolates analgesic activity by *in-vivo* hot plate method using mice. *Trikatu churna* was found to possess promising antimicrobial (non-specific) activity on *Escherichia coli*, *Staphylococcus aureus*, *Aspergillus niger* and *Mucor* species by *in-vitro* agar well diffusion method (Reddy and Seetharam 2009).

The aqueous, ethanol, methanol and acetone extracts of *Piper longum*, *Piper nigrum* and *Zingiber officinale* fruits and *Trikatu churna* were prepared and antibacterial activities were tested by disc diffusion method against enteric bacterial pathogens. The extracts were found antibacterial to all *Staphylococcus aureus*, *Pseudomonas aeruginosa*, *Proteus vulgaris*, *Staphylococcus epidermidis*, *Salmonella typhi*, *Salmonella typhimurium* and *Enterobacter aerogenes*. *Trikatu churna* exhibited potent antibacterial activity; this might be due to the multifunctional effect of all the three plant ingredients (Dahikar et al. 2010).

Anticancer: 20-Methylcholanthrene was used to induce tumor in albino mice. Individual treatment with Mercaptopurine (5 mg/kg) and *Trikatu* (100 mg/kg) significantly restored the altered haematological and antioxidant parameters to normal values. Even Mercaptopurine (2.5 mg/kg) at its sub therapeutic dose showed equivalent effects as that of therapeutic dose of Mercaptopurine (5 mg/kg) when it was co-administered along with *Trikatu* compared to the positive tumor control group (Dsouza et al. 2013).

Hepatoprotective: The ethanol extract of *Trikatu Churna* at an oral dose of 150 mg/kg exhibited a significant protective effect by lowering serum levels of glutamic oxaloacetic transaminase, glutamic pyruvic transaminase, alkaline phosphatase and total bilirubin in rats by inducing liver damage with carbon tetrachloride. Liv 52 syrup was used as positive control (Kumar and Mishra 2004).

Hypolipidemic: *Trikatu* was fed to normal and cholesterol fed male *Rattus norvegicus* to study its efficacy as a hypolipidemic activity. *Trikatu* reduced triglycerides and LDL cholesterol and increased HDL cholesterol thereby reducing the risk of hyperlipidemia and atherosclerosis (Sivakumar et al. 2004).

Toxicity

Acute toxicity: *Trikatu* at 2,000 mg/kg body weight once orally was well tolerated by the experimental animals (both male and female) and no changes were observed in mortality, morbidity, gross pathology, gain in weight, vital organ weight, haematological, biochemical parameters, serum lipid profile and tissue biochemical.

Sub-acute toxicity: In sub-acute experiment, *Trikatu* was administered at 5, 50 and 300 mg/kg body weight once daily for 28 days in female Charles Foster rats. There was significant increase in low density lipoprotein cholesterol level at 50 and 300 mg/kg body weight, decrease in high density lipoprotein cholesterol level at 300 mg/kg body weight, increase in SGPT activity at 50 mg/kg body weight and decrease in WBC count at 300 mg/kg body weight on 28(th) day post treatment (Debabrata et al. 2009).

Clinical Study

Chronic rhinitis: In a study, treatment with *Trikatu yoga* on 42 patients diagnosed with chronic rhinitis, fresh cases got greater relief as compared to older cases. *Trikatu yoga* prevented recurrence of chronic rhinitis in new cases and provided resistance against seasonal attack (Sridhar 2001).

Triphala Ghrita

Preclinical pharmacology

Anticataract: Administration of *Triphala ghrita* at a dose of 216 mg/200 g, 1080 mg/200 g and 2160 mg/200 g of rat orally offered significant dose dependent protection against galactose induced cataract and delayed the onset and progression of cataract. IDose of 1080 mg of *Triphala ghrita* offered significant protection against delaying the onset and progression of cataract in comparison to other doses (Mahajan et al. 2011).

Triphala Guggulu

Pharmaceutical and analytical evaluation: *Triphala guggul* kalpa tablet, described in sharangdharsanhita and containing guggul and triphala powder, was used as a model drug. Preliminary experiments on marketed *triphala guggul kalpa* tablets exhibited delayed *in vitro* disintegration that indicated probable delayed *in vivo* disintegration. Tablets, prepared by direct compression method, complied with the hardness and disintegration tests, whereas tablets prepared by Ayurvedic text methods failed (Savarikar et al. 2011).

Clinical study

Bleeding piles: Present study was carried out using a combination of *Apamarga Kshara Basti* and *Triphala guggulu*. The results of the clinical assessment of the indigenous formulation on 129 patients with bleeding piles are reported in this paper; 55 patients of a total of 129 showed marked relief (Mehra et al. 2011).

Hyaluronidase and collagenase inhibitory: Aqueous and hydro-alcoholic extracts of *Triphala shodith guggulu* showed weak but dose-dependent inhibition of hyaluronidase activity. In contrast, the TG formulation was 50 times more potent than the *Triphala shodith guggulu* extract with respect to hyaluronidase inhibitory activity. Hydro-alcoholic extracts of the *Triphala guggulu* were four times more potent than *Triphala shodith guggulu* with respect to collagenase inhibitory activity (Sumantran et al. 2007).

Triphala Eye Drops

Clinical Study

Computer vision syndrome: In a clinical study, 151 patients were registered, out of whom 141 completed the treatment. In Group A, 45 patients had been prescribed *Triphala* eye drops; in Group B, 53 patients had been prescribed the *Triphala* eye drops and *Saptamrita Lauha* tablets internally, and in Group C, 43 patients had been prescribed the placebo eye drops and placebo tablets. In total, marked improvement was observed in 48.89, 54.71 and 06.98% patients in groups A, B and C, respectively (Gangamma and Rajagopala 2010).

Triphala Mashi

Preclinical pharmacology

Anti-diarrhoeal: Aqueous and alcoholic extracts of *Triphala Mashi* at various doses 200, 400 and 800 mg/kg displayed remarkable anti-diarrhoeal activity as evidenced by a significant increase in first defecation time, cumulative fecal weight and intestinal transit time (Biradar et al. 2007).

Antibacterial: *Triphala Mashi* containing phenolic compounds, tannins exhibited comparable antimicrobial activity in relation to Triphala against all the microorganisms tested. It inhibits the dose-dependent growth of Gram-positive and Gram-negative bacteria (Biradar et al. 2008).

Toxicity: Aqueous and alcoholic extracts of *Triphala Mashi* were considered safe up to a dose of 1750 mg/kg when evaluated for acute oral toxicity in accordance with the OECD guidelines (Biradar et al. 2007).

Trivanga Bhasma

Formulation, characterization and comparative evaluation: The X-ray Diffraction analysis of *Trivanga bhasma* (TB1) prepared as per Ayurvedic formulary and marketed *Trivanga bhasma* (TB2) exhibited crystalline nature and nano-sized particles by Scherrer's equation. In SEM studies, lead, zinc and tin oxides show well-defined plate like structures while TB1 showed spongy, relatively compact microcrystalline aggregates with loss of grain boundaries. AFM analysis confirmed the spherical morphology of TB1 and TB2 with an average particle size of 500 nm (Rasheed et al. 2014).

Standardisation: The quantitative estimation of various metallic constituents of *Trivanga Bhasma*, was carried out which provided following results (Sharma and Singh 1987).

Table: Analytical results of *Trivanga Bhasma*.

Lead	Tin	Zinc	Iron	Aluminium
24.4	24.26	24.4	4.13	0.79

Toxicity: Group I served as control and was given vehicle (honey:water in 2:3 ratio) Group II, III, and IV received *Trivanga Bhasma* at 7.8, 39.5, and 78 mg/kg body weight for 90 consecutive days. *Trivanga Bhasma* was found to be safe. No significant clinical signs were noted in all groups studied. No major alterations were observed during histopathological evaluation. Hence, dose rate of 78 mg/kg body weight was established as NOAEL (Pallavi et al. 1987).

Vachadhatryadi Avaleha

Preclinical pharmacology

Immunomodulatory: *Vachadhatryadi Avaleha* was administered at the dose of 900 mg/kg and parameters like hemagglutination titer, ponderal changes, histopathology of immunological organs and immunological paw edema were recorded. *Vacha Dhatryadi Avaleha* significantly enhanced antibody formation and moderately suppressed the immunological edema (Rajagopala et al. 2011).

Vaikranta Bhasma

Physico-chemical study: Scanning electron microscopy, energy dispersive X-ray analysis, Fourier transform infrared spectrometry and inductively coupled plasma spectrometry show *Vaikranta bhasma* to be a multi-mineral compound, which contains iron and silica as major constituents and others are present as trace elements (Tripathi et al. 2013).

Vaishvanar Churna

Clinical study

Rheumatoid arthritis: *Vaishvanar Churna* was given to the 50 *rheumatoid arthritis* patients in the dose of 3 gm twice a day in the powder form for eight weeks with warm water. During the study, *Vaishvanara Churna* was capable of producing significant symptomatic improvement and can alleviate the disease in an earlier stage.

Vanga Bhasma (Incinerated Tin)

XRD analysis: XRD peaks of *Vanga Bhasma* are identified to be as Tindioxide (SnO_2).

Preclinical pharmacology

Testicular regenerative potential: *Vanga Bhasma* in reference is found to have testicular regenerative potential on cadmium induced testicular degeneration in albino rats, when administered orally (Nagaraju et al. 1985).

Toxicity: But for local irritation, no significant toxicity attributable to *Vanga bhasma* has been observed even in eight times higher dose than the therapeutic dose, on exposure to the drug for ten days (Nagaraju et al. 1984).

Clinical study

Diabetes mellitus: In a clinical study, *Vanga bhasma* was administered to 30 diagnosed patients of *diabetes mellitus* and a favourable effect was reported (Lagad and Ingole 2009).

Vardhamana Pippali Rasayana

Preclinical pharmacology

Rheumatoid Arthritis: 73 patients were given *Vardhamana Pippali Rasayana* for 15 days. Patients with any other acute or chronic systemic illness or infection were excluded from the study. The observations and results obtained were analyzed statistically applying the "t" test. All the patients experienced up to 50% relief from the signs and symptoms of rheumatoid arthritis after the therapy (Soni et al. 2011).

Varunadi Loha

Preclinical pharmacology

Nephroprotective: In the control group, the results were not significant as the rats were non diseased and kept on normal diet. In the disease control group, results were highly significant in serum creatinine and serum urea levels as the levels were raised. In the curative group, results were significant in serum levels as there was moderate decrease in serum creatinine and serum urea levels after the treatment with *Varunadi loha*.

Varunshigru Ghan Vati

Clinical study

Urinary tract infection: Thirty patients were selected for this study and divided randomly in two Groups I (Treated by *Varunshigru Ghan Vati*) and II (Treated by antibiotic as per culture and sensitivity of urine). Patients were called for five follow ups, first three at the interval of 7 days and then two follow up at the interval of 15 days. The results prove treatment was equivalent to antibiotics in most situations and better results in some circumstances.

Vasavaleha

Stability Study: Physic-chemical parameters and residual drug content of *Vasavaleha* (AVB, AVB-1 and AVB-2) was measured at accelerated temperature. No change was noticed in color, odor and taste of AVB up to storage of 180 days. In pH, slight changes were noticed in all the formulations. Moisture content at 0 day of AVB was 3.75% which reduced after 60 days to 3.74% and after 120 days it was 3.72% while after 180 days it was found the same 3.72%. The percent residual drug content of prepared *vasavaleha* was found 94.15% and 92.11%, 92.15% for AVB-1 and AVB-2 after 180 days at 40 ± 1°C while 98.05%, 98.21%, 98.08% was estimated after 60 days at 40 ± 1°C respectively for AVB, AVB-1 and AVB-2 (Ahirwar 2013).

Vasaguduchyadi Kwatha

Acute toxicity: Acute toxicity test was evaluated as per OECD 425 guidelines with 5,000 mg/kg as limit test in Wistar strain albino rats. Test formulations were administered to overnight fasted animals and parameters like body weight, behavioral changes, and mortality were assessed for 14 days. Hematological and biochemical parameters were assessed on the 14th day. Results showed no significant changes in terms of behavioral changes, mortality, and body weight. However, increase in blood urea level was observed (Kotecha et al. 2013).

Vara Asanadi Kwatha

Pharmacognostical and phytochemical evaluation: Organoleptic features of coarse powder made out of the crude drugs were within the standard range. Specific gravity of the decoction was 1.0185 and pH was 5.5. Total solid content present in the Kwatha was 4.525% w/v, total ash 0.949% w/v, and acid insoluble ash was 0.052% w/v. Iron assay showed the presence of Fe(2)O(3) as 0.065% w/v. Qualitative scrutiny demonstrated the presence of flavonoids and tannins (Ramachandran 2012).

Varatika bhasma (Karpardika bhasma)

Chemical standardisation: SEM and XRD analysis further confirmed the occurrence of nano crystalline compounds in the final product. Chemically it contains tannins and flavones (Rajalakshmi and Brindha 2010). A comparative study of *Kapardika* powder and four commercial samples of *Kapardika bhasma* are carried out using modern instrumental techniques. Four samples of *Kapardika bhasma* resemble each other with minor differences due to presence of a small quantity of water, calcium hydroxide and organic matter (Dhamal et al. 2012).

Pre-clinical pharmacology

Gastroprotective: This study reported efficacy of *Varatika bhasma* in ulcer protective activity with standard control as Sucralfate in aspirin induced ulcer (Bhatt 2013).

Vasant Kusumakara Rasa

Clinical study

Diabetic neuropathy and *retinopathy*: A clinical evaluation of *Vasant kusumakara rasa* in complications of diabetes mellitus viz diabetic neuropathy and retinopathy has been done (Tamboli 2001).

Vatagajankusa rasa

Clinical study

Post polio paresis: A study reported effect of *Vatagajankusa rasa* on post polio paresis (Sharma 1997).

Vatari Guggulu

Clinical study

Rheumatoid arthritis: Twenty two patients suffering from rheumatoid arthritis were given *Vatari Guggulu* and *Maharasnadi Quatha* internally for 3 to 6 weeks. The results were assessed in terms of clinical recovery and functional improvement. Relief from pain, oedema and fever was statistically significant (P < 0.01) and fall in ESR levels was highly significant (P < 0.001) (Swamy and Bhattathiri 1998).

For this study, 118 patients of *Amavata* were randomly divided into two groups. The patients in group A (50 patients) were given *Matra Basti* with *Brihat Saindhavadi Taila* along with *Vatari Guggulu*; the patients in group B (53 patients) were given only *Vatari Guggulu*. All the patients responded favorably to the treatment in both the groups; however, patients treated with *Matra Basti* had better relief in most of the cardinal signs and symptoms of the disease (Khagram et al. 2010).

Vayasthapana Rasayana

Chemical composition: The total phenolic content (gallic acid equivalent) of the *Vayasthapana Rasayana* is 8.3 mg per g of dry mass.

Pre-clinical pharmacology

Antioxidant: Methanol extracts of *Vayasthapana Rasayana* were studied for *in vitro* total antioxidant activity. The formulation has shown 94% at 0.1 mg/ml DPPH free-radical scavenging activity as against 84% at 0.1 mg/ml for standard ascorbic acid (IC_{50} value 5.51 µg/ml for VRF and 39 µg/ml for standard). ABTS radical scavenging activity of *Vayasthapana Rasayana* was 69.55 ± 0.21% at 100 µg/ml concentration with an IC50 value of 69.87 µg/ml (Mukherjee et al. 2011).

Veerataru Kwatha

Pre-clinical pharmacology

Diuretic: Randomly selected animals were divided into three groups of six animals each. *Veerataru Kwatha* was administered orally at a dose of 5.4 and 10.8 ml/kg. Parameters like volume of urine, pH of urine and urinary electrolyte concentrations like sodium, potassium and chloride were studied. *Veerataru Kwatha* increased the urine output in a dose-dependent manner (Patel et al. 2011).

Vibhitakyadi Churna

Clinical study

Bronchitis: A clinical study reported antitussive and antiasthmatic effects of *Vibhitakyadi churna* in the cases of *kasa-swasa* (Trivedi, Nesamony and Sharma 1982).

Vidangadi churna

Clinical study

Worm infestation: *Vidangadi churna* was tested using adult earthworm *Pheritima posthuma* against Piperazine citrate (15 mg/ml) and albendazole (20 mg/ml) as standard references and normal saline as control. The time to achieve paralysis of the worms was determined. *Vidangadi churna* produced a potent anthelminthic activity against the *P. posthuma* when compared with reference standards (p < 0.001) (Sajith et al. 2013).

Vidarikandadi Yog

Clinical study

Performance enhancing: In a randomized double blind placebo controlled study, anabolic effect of *Vidarikandadi Yog* was studied. The students were randomly divided into two groups. Group A (*Vidarikandadi Yog*) comprising of 38 and Group B (placebo) of 34 students. *Vidarikandadi Yog* was given in the dose of 200 mg/kg/day in two divided doses for two months with milk and follow up was conducted fortnightly. Significant results for weight and chest circumference were noticed, whereas highly significant results were obtained for muscular strength and endurance assessment parameters (Ingle et al. 2013).

Virya Sthambhak Vati

Clinical study

Aphrodisaic: *Virya Sthambhak Vati* was effective both clinically and statistically in terms of increase in the coital duration, reduction in mental stress and improvement in penile rigidity. The drug showed no adverse effect and therefore can be continued for longer duration also (Randhir and Sharma 2010).

Vyoshadi Guggulu

Clinical study

Dyslipidemia: *Vyoshadi Guggulu* (1 gm thrice in day) and *Haritaki Churna* (3 gm twice in day) was selected for the management of dyslipidemia in the present study & given for duration of 12 weeks. Total 47 out of 53 patients were completed. After completion of treatment there was remarkable percentage improvement in subjective criteria like breathlessness, parasthesia, confusion, & fatigue (Ragvani et al. 2013).

Yashada Bhasma (Incinerated Zinc or Zinc Calx)

Analytical Study: Spectroscopic analysis of *Yashada bhasma* confirms presence of Zinc Oxide (Thakur et al. 1986). Scanning Electron Microscopy revealed the amorphous nature of the bhasma with particle size range 5–20 μm. Inductively Coupled Plasma Atomic Emission Spectroscopy showed the presence of Zinc in major portion (95.08 ppm) and other elements like Sn (0.27), Pb (0.14), Fe (1.69), Ca (1.82), Mg (1.00), Cu, Co and Mn < 0.5 ppm in the final product (Santhosh et al. 2013).

Preclinical pharmacology

Antiacne: The present study revealed the inhibitory effect of *Yashada Bhasma*, *Tankana* and combination of both on *P. acne* growth and inflammation. Clinical studies have suggested the anti-acne benefits of formulations containing *Yashada Bhasma* and *Tankana*. The findings obtained from the present *in vitro* studies provide evidence to support the mechanism of anti-acne properties of *Yashada Bhasma* and *Tankana*.

Antidiabetic: A study reported antidiabetic action of *Yashada bhasma* instreptozotocin-induced diabetes (Rao et al. 1997).

Antioxidant: The study was carried out with the drug, in suspension form, of three different concentrations, i.e., 1%, 2% and 5% and the duration of administration as one day, two days and four days for each suspension. The results showed that for preventing lipid peroxidation concentration of 2% *Yashada bhasma* was sufficient and in four days study the results were better than standard drug (Vit C). SOD, GSH and CAT study revealed better results with 5% *Yashada bhasma* suspension (Santhosh et al. 2013).

Hypoglycaemic: Hypoglycaemic activity of *Shilajatu* and *Yashada Bhasma* was determined in normoglycaemic and alloxanised albino rats. Significant hypoglycaemic activity was observed in the treated groups. However, the activity was found to be comparatively milder in *Shilajatu* treated rats compared to *Yashada Bhasma* treated rats (Bharati et al. 1998).

Immunomodulator: *Yashada bhasma* was subjected to screening of immuno modulatory effect using three healthy adult human blood samples and with three different drug concentrations, i.e., 1%, 2% and 5%. The parameters used were Nitro blue tetrazolium assay, Phagocytosis, Candidacidal assay and Chemotaxis. 5% drug suspension showed significant results in all the four parameters (Bhojashettar et al. 2013).

Toxicity: A comprehensive safety study of *Yashada Bhasma* in three dosage forms on various parameters such as hematological, biochemical and histo-pathological, etc. The safety profile of *Yashada bhasma* shows it to be quite safe (Joshi et al. 2008).

Yavanyrka

Physico-chemical characteristics: Thymol, pinene, p-cymene, dipentene and terpinene have been separated chromatographically (Dhyani and Baxi 1985).

Yavakshara (Alkali preparation of Barley)

Clinical study

Benign Prostatic Hypertrophy: The control study was carried out in two different groups. Each group made of 30 patients. In trial group 30 patients were treated with *Yavakshara*. In the control group 30 patients were treated with capsule Pyginal (containing *Pygeum africanum*). Most of the patients having associate symptoms, i.e., constipation and *Yavakshara* showed marked improvement for constipation while pyginal does not show any relief in constipation (Borkar et al. 2013).

Nephrolithiasis: A comparative study was conducted on 20 patients. Ten patients of Group I received *Yava Kshara* and the remaining 10 patients were given an indigenous drug called GCP Compound for a period of 60 days. Statistical analysis showed that P-value of less than 0.05 was seen in Group I from 0 to 60 days course. There was no significant difference observed in the stone size in Group I when the actual value and calculated value were compared (Chakradhar and Datatreya 2013).

Yavaksharadi vati

Tonsillitis: A clinical study reorted efficacy of *Yavaksharadi vati* and *Panchavalkal kwath* in the management of tonsillitis (Ahuja 2014).

Yograj Guggul

Preclinical pharmacology

Phase I tolerability study: Eight male students volunteered for the Phase 1 study. They were divided into four groups of two each for different dose schedules. A careful history, thorough physical examination and laboratory investigations were carried out to exclude any systemic disease. The age of the volunteers ranged from 22 to 28 years. The body weight ranged from 45 to 63 kg.

The general tolerability of *Yogaraj-guggulu* was good in volunteers even at the dose of 9 gm per day. Out of three volunteers who developed diarrhoea, one had ova of *Trichuris trichiura* and another had ova of tape worm (Antarkar et al. 1984).

Zahr Mohra (Ophite or Serpentine)

Pre-clinical pharmacology

Antioxidant: *Zahr Mohra* produces a significant increase in Glutathione reductase activity, in comparison of the untreated, stressed animals, which was not significantly different from the Glutathione reductase level in a-tocopherol-treated animals as well as in the nonstressed animals (Ali et al. 2009).

Bibliography
(Part B)

Abdullah-Al-Mamun M, Huque A, Biswas S, Bhuiyan R, Harun Ur Rashid M. Toxicological studies of *Karpura Rasa* on the basis of lipid profile, liver function and kidney function in rat plasma after chronic administration. *J Chem Pharma Res* 2014; 6: 122–6.

Abhishek J Joshi, Aparna K, Rajagopala S, Kalpana S Patel, Harisha CR, Vinay J Shukla. A preliminary pharmacognostical and pharmaceutical evaluation of *Bala Chaturbhadra Avaleha*. *Ann Ayur Med* 2014; 3: 20–8.

Acharya SB, Frotan MH, Goel RK, Tripathi SK, Das PK. Pharmacological actions of *Shilajit*. *Indian J Exp Biol* 1988; 26: 775–777.

Achliya GS, Wadodkar SG, Dorle AK. Evaluation of hepatoprotective effect of *Amalkadi Ghrita* against carbon tetrachloride-induced hepatic damage in rats. *J Ethnopharmacol* 2004; 90: 229–32.

Agarwal A, Girija PV. Physico-chemical standardization of *Narcha churna* – an Ayurvedic formulation. *Bull Medico-Ethno-Bot Res* 2005; 26: 47–53.

Agrawal RG, Joshi PC, Pandey MJ, Pandey G. Studies on standardisation of *Tribhuvan Kirti Rasa*. *Anc Sci Life* 1996; 15.

Ahir Y, Ila tanna, Ravishankar B, Chandola HM. Evaluation of clinical effect of *Kushmandadi Ghrita* in Generalized Anxiety Disorder. *Indian J Trad Knowl* 2011; 10: 239–46.

Ahirwar B. Evaluation of stability study of Ayurvedic formulation *Vasavaleha*. *Asian J Pharm Res* 2013; 3: 1–4.

Ahirwar B, Ahirwar D, Alpana R. Antihistaminic effect of *Sitopaladi Churna* extract. *Res J Pharm Technol* 2008; 1: 89–92.

Ahuja DK. Clinical study on *Yavaksharadi vati* and *Panchavalkal kwath* in the management of tundikeri w.s.r. to tonsillitis. *J Ayur Holis Med* 2014; 2: 23–31.

Akhtar B, Mahto RR, Dave AR, Shukla VD. Clinical study on *Sandhigata Vata* w.s.r. to Osteoarthritis and its management by *Panchatikta Ghrita Guggulu*. *Ayu* 2010; 31: 53–7.

Alam M, Rukmani B, Chelladurai V, Usman Ali S, Purushothaman KK. Chemical examination of commercial *Gomutra Shilajit*. *Bull Medico-Ethno-Bot Res* 1980a; 1: 525–9.

Alam M, Sathiavasan K, Dasan KKS, Purushothaman KK. Standardisation of *Dasamula Taila*. *J Res Ayur Siddh* 1980b; 2: 79–84.

Alam M, Sathiavasan K, Varadarajan TV, Dasan KKS, Purushothaman KK. A colorimetric method to standardise *chandraprabha vati*. *J Res Ayur Siddha* 1982; 3: 216–20.

Alam M, Dasan KK, Ramar C, Ali SU, Purushothaman KK. Experimental studies on the fermentation in asavas and aristas part-1 *Draksarista*. *Anc Sci Life* 1983a; 2: 148–52.

Alam M, Rukumani B, Usman Ali S, Purushothaman KK. Studies on authentic *Shilajit* samples. *J Drug Res Ayur Sidd* 1983b; 4: 54–61.

Alam M, Dasan KK, Joy S, Purushothaman KK. Chemical, microbiological and comparative and fermentation studies on *dasamularishta*. *Anc Sci Life* 1984a; 4: 123–6.

Alam M, Rukmani B, Shanmughadasan KK, Purushothaman KK. Effect of time on the fermentation and storage of *candanasava*. *Anc Sci Life* 1984b; 4: 51–5.

Alam M, Dasan KKS, Rukmani B, Purushothaman KK. Studies on the standardisation of *Drakshadi vati*. *Bull Medico-Ethno-Bot Res* 1985; 6: 133–40.

Alam M, Dasan KK, Rukmani B, Veni H, Purushothaman RG, Purushothaman KK. Experimental studies on the fermentation of *aravindasava*. *Anc Sci Life* 1986; 5: 243–6.

Alam M, Dasan KK, Joy S, Purushothaman KK. Comparative and fermentation standardisation studies on *dasamularishta*. *Anc Sci Life* 1988a; 8: 68–70.
Alam M, Dasan KKS, Joy S, Bhima Rao R. Studies on the standardization of *Hinguadi taila*. *J Res Ayur Sidd* 1988b; 9: 146–9.
Alam M, Dasan KKS, Joy S, Purushothaman KK. Studies on the standardization of *Mahalaksadi Taila*. *Anc Sci Life* 1989a; 9: 28–30.
Alam M, Rukmani B, Meenakshi N, Dasan KKS, Bhima Rao R. Standardisation studies on some *Dasamula* containing formulations. *J Res Ayur Siddh* 1989b; 14: 68–73.
Alam M, Dasan KKS, Meenakshi N, Bhima Rao R. Studies on the standardization of curnas Part-II *Talisadya Churna*. *Anc Sci Life* 1991; 11: 46–9.
Alam M, Dasan KKS, Bhima Rao R. Standardisation of *Karpurasava*. *Anc Sci Life* 1994; 14: 49–52.
Alam M, Shanmuga Dasan KK, Thomas S, Suganthan J. Anti-inflammatory potential of *Balarishta* and *Dhanvantara Gutika* in albino rats. *Anc Sci Life* 1998; 17: 305–12.
Al-Amin Md. M, Hossain S, Shohel M, Mahmud S. *Arkadi Kvatha Churna* and its study on different parameters in rats. *Int Ayur Med J* 2013; 1: 1–7.
Ali SZ, Khan NA, Amin KMU, Rizvi SJ, Zaidi SMK, Banu N. The possible role of glutathione reductase in the glutathione mediated antioxidant activity of *Zahr Mohra* (Serpentine) – An experimental study. *Hippocratic J Unani Med* 2009; 4: 35–40.
Anandjiwala S, Bagul MS, Parabia M, Rajani M. Evaluation of free radical scavenging activity of an ayurvedic formulation, *Panchvalkala*. *Indian J Pharm Sci* 2008; 70: 31–5.
Anonymous. Effect of *Ashokarishta* and *Musalikhadiradikwatha* in menorrhagia. Government Ayurveda College, Trivandrum, 2002.
Antarkar DS, Pande RR, Athavale AV, Shubhangi RR, Saoji SR, Shah KN, Jakhmola AT, Vaidya A B. Phase I tolerability study of *Yogaraj-guggulu* – a popular Ayurvedic drug. *J Postgrad Med* 1984; 30: 111.
Anuradha D, Anish N, Naidu ML, Srinivasulu M. Clinical evaluation of efficacy of *Pippali* with *Rasnasaptaka* in management of *Amavata*. E-Journal *Rasamruta* 2013; 5: 1–5.
Apaturkar N, Pimpalkar P, Pal R, Raut S, Lakhapati AM. Role of *Apamarga Kshara* with *Apamarga Kshara Tailam* on *dushtavrana* with reference to infected wound-a case. E-Journal *Rasamruta* 2013; 5: 1–8.
Arora P, Ansari SH. Quality standard parameters of anti-asthmatic ayurvedic formulation *"Kanakasava"*. *Int J Pharmacog Phytochem Res* 2013; 6: 85.
Arun Raj GR, Nikhil DV, Shailaja U, Rao Prasanna N, Ajayan S. Clinical study on efficacy of local application with *Jyotishmati-apamarga kshara taila* in the management of *shvitra* (vitiligo) in children. *Univ J Pharm* 2013; 2: 89–92.
Arya N, Anil Kumar E, Maheswar T, Madhavi N. Standardization of *Surya Shekhara Ras*: A Kupi Pakwa Rasayana. *Int J Res Ayur Pharm* 2013; 4: 670–5.
Ashok BK, Ravishankar B, Prajapati PK, Bhat SD. Antipyretic activity of *Guduchi Ghrita* formulations in albino rats. *Ayu* 2010; 31: 367–70.
Aswatha Ram HN, Kaushik U, Lachake P, Shreedhara CS. Standardisation of *Avipattikar Churna* – A polyherbal formulation. *Phcog Res* 2009; 1: 224–7.
Aswatha Ram HN, Sriwastava NK, Makhija IK, Shreedhara CS. Anti-inflammatory activity of *Ajmodadi Churna* extract against acute inflammation in rats. *Ayurveda Integr Med* 2012; 3: 33–7.
Avula B, Wang YH, Rumalla CS, Chittiboyina AG, Srivastava A, Srivastava K, Padhi MM, Babu R, Sharma SK, Khan IA. Chemical analysis of an Ayurvedic herbo-metallic preparation containing mercury: Speciation of mercury. *Planta Med* 2011; 77.
Babbar AK. Study on clinical efficacy of *Sanjivani vati* and *Lashunadi vati* in management of diarrhea predominant IBS: A pilot study. *J Ayur Hol Med* 2014; 2: 13–22.
Babu G, Bhuyan GC, Prasad GP, Swamy GK. The clinical effect of *Shunthi Guggulu* and *Godanti* in the management of *Amavata*. *J Res Ayur Siddh* 2009; 30: 39–50.
Bagul MS, Srinivasa H, Kanaki NS, Rajani M. Anti-inflammatory activity of two Ayurvedic formulations containing guggul. *Indian J Pharmacol* 2005a; 37: 399–400.
Bagul MS, Kanaki NS, Rajani M. Evaluation of free radical scavenging properties of two classical polyherbal formulations. *Indian J Exp Biol* 2005b; 43: 732–6.
Bajaj S, Vohora SB. Analgesic activity of gold preparations used in Ayurveda & Unani-Tibb. *Indian J Med Res* 1998; 108: 104–11.
Bajaj S, Vohora SB. Anti-cataleptic, anti-anxiety and anti-depressant activity of gold preparations used in Indian systems of medicine. *Ind J Pharmacol* 2000; 32: 339–46.

Bajaj S, Ahmad I, Raisuddin S, Vohora SB. Augmentation of non-specific immunity in mice by gold preparations used in traditional systems of medicine. *Indian J Med Res* 2001; 113: 192–6.

Bandari S, Bhadra Dev P, Murthy PHC. HPTLC fingerprinting in the standardization of *Panchavaktra Rasa*: A herbo-mineral preparation. *Int J Ayur Med* 2011; 2.

Bandari S, Bhadra Dev P, Murthy PHC. Clinical evaluation of *Panchavaktra Rasa* in the management of *Amavata* (Rheumatoid Arthritis). *Int J Ayur Med* 2012; 1.

Bansal C. Comparative study on the effect of *Saptamrita Lauha* and Yoga therapy in myopia. *Ayu* 2014; 35: 22–7.

Bansal N, Parle M. Beneficial effect of *Chyavanprash* on cognitive function in aged mice. *Pharm Biol* 2011; 49: 2–8.

Bapat V Rao. Epilepsy. *Sach Ayur* 1982; 34: 485–90.

Baragi PC, Baragi UC, Bhat S, Prajapati PK. Physico-chemical profile of *Puga Khanda*: A preliminary study. *Ayu* 2014; 35: 103–7.

Baragi UC, Baragi PC, Vyas MK, Shukla VJ. Standardization and quality control parameters of *Dashanga Kwatha Ghana* tablet: An Ayurvedic formulation. *Int J Ayurveda Res* 2011; 2: 42–7.

Baria J, Gupta SK, Bhuyan C. Clinical study of *Manjishthadi Ghrita* in *vrana ropana*. *Ayu* 2011; 32: 95–9.

Bavadekar SS, Tagare AR, Kakar T, Bakare SC, Ingole RK. Preparation and physicochemical analysis of *Mandoor parpati*. *Int Res J Pharm* 2013; 4: 257–9.

Behera B, Yadav D, Sharma MC. Effect of an herbal formulation (*Indrayanadi Yog*) on blood glucose level. *Int Res J Biol Sci* 2013; 2: 67–71.

Bhakti C, Rajagopala M, Shah AK, Bavalatti N. A clinical evaluation of *Haridra khanda* and *Pippalyadi taila nasya* on *pratishyaya* (Allergic rhinitis). *AYU* 2009; 30: 188–93.

Bhalerao S, Munshi R, Nesari T, Shah H. Evaluation of *Brahmi ghrita* in children suffering from attention deficit hyperactivity disorder. *Anc Sci Life* 2013; 33: 123–30.

Bhalodia SG, Bhuyan C, Gupta SK, Dudhamal TS. *Gokshuradi Vati* and *Dhanyaka-Gokshura Ghrita Matra Basti* in the management of Benign Prostatic Hyperplasia. *Ayu* 2012; 33: 547–51.

Bharati, Chansouria JPM, Singh RH. Studies on hypoglycemic activity of *Shilajatu* and *Yashada Bhasma*. *J Res Ayur Siddh* 1998; 19: 64–7.

Bhargava Sl. Evaluation of antimicrobial potential of *Sudarshan churna*: a polyherbal formulation. *Iranian J Pharmacol Ther* 2008; 7: 185–7.

Bharti A, Ahirwar D, Alpana R. Antihistaminic effect of *Sitopaladi Churna* extract. *Res J Pharm Technol* 2008; 1: 12–6.

Bhatia B, Kale PG. Analytical evaluation of an Ayurvedic formulation – *Abhraka Bhasma*. *Int J Pharm Sci Rev Res* 2013; 23: 17–23.

Bhatia BS, Kale PG, Daoo JV, Panchal PP. *Abhraka Bhasma* treatment ameliorates proliferation of germinal epithelium after heat exposure in rats. *Anc Sci Life* 2012; 31: 171–80.

Bhatt P. Evaluation of *Varatika bhasma* for its ulcer protective effect on albino rats. *Anc Sci Life* 2013; 32: 15.

Bhattacharya SK. Activity of *Shilajit* on alloxan-induced hyperglycaemia in rats. *Fitoterapia* 1995a; 66: 328–32.

Bhattacharya SK. *Shilajit* attenuates streptozotocin induced diabetes mellitus and decrease in pancreatic is let superoxide dismutase activity in rats. *Phytotherapy Res* 1995b; 9: 41–4.

Bhaumik S, Chattapadhay S, Ghosal S. Effects of *Shilajit* on mouse peritoneal macrophages. *Phytother Res* 1993; 7: 425–427.

Bhojashettar S, Jadar PG, Nageswara R. Experimental evaluation of immunomodulatory effect of *Yashada bhasma* (Incinerated Zinc). *Int Ayur Med J* 2013; 1: 1–9.

Bhondave PD, Devarshi PP, Mahadik KR, Harsulkar AM. *Ashwagandharishta* prepared using yeast consortium from *Woodfordia fruticosa* flowers exhibit hepatoprotective effect on CCl4 induced liver damage in Wistar rats. *J Ethnopharmacol* 2014; 151: 183–90.

Bhuyan GC, Srikanth N. Multi centric Open Observational study on Clinical Safety and Efficacy of *Dhatri lauha* – (A classical Ayurvedic Formulation) in the management of Iron Deficiency Anemia (Pandu). Proceedings of 4th World Ayurveda Congress; Bangaluru; India 9–13th December 2010 p.p 437-348 4th World Ayurveda Congress; Bangaluru; India 9–13th December 2010; 01/2010.

Biradar YS, Singh R, Sharma K, Dhalwal K, Bodhankar SL, Khandelwal KR. Evaluation of anti-diarrhoeal property and acute toxicity of *Triphala Mashi*, an Ayurvedic formulation. *J Herb Pharmacother* 2007; 7: 203–12.

Biradar YS, Jagatap S, Khandelwal KR, Singhania SS. Exploring of antimicrobial activity of *TriphalaMashi*—an *Ayurvedic* formulation. *eCAM* 2008; 5: 107–13.

Biswas TK, Pandit S, Mondal S, Biswas SK, Jana U, Ghosh T, Tripathi PC, Debnath PK, Auddy RG, Auddy B. Clinical evaluation of spermatogenic activity of processed Shilajit in oligospermia. *Andrologia* 2010; 42: 48–56.

Bordia A, Verma SK, Mewara OP, Kaushik SK. The effect of Chyavanprash on blood lipids, fibrinolysis, platelet adhesiveness and vitamin C levels in man. *Nagarjun* 1981; 14: 44–8.

Borkar KM, Shekokar AK, Singh AK. A control study of *Yavakshara* in the management of Benign Prostatic Hyperplasia (BPH) W.S.R. to *Vatastheela*. *Int J Ayur Pharma Res* 2013; 1: 60–5.

Bulbul L, Choudhuri MSK. Toxicological studies of *Asmarihara kasaya curna* – an ayurvedic formulation on liver function parameters after chronic administration. *Ind Amr J Pharm Res* 2013; 3: 3734–41.

Buwa S, Patil S, Kulkarni PH, Kanase A. Hepatoprotective action of *Abhrak bhasma*, an ayurvedic drug in albino rats against hepatitis induced by CCl4. *Indian J Exp Biol* 2001; 39: 1022–7.

Chakraborty B. Study on the effect of chronic administration of *Khadirarista* on different hematologic parameters in male Sprague-Dawley rats. Conference: International Conference on Updates on Natural Products in Medicine and Healthcare System, Bangladesh. http://www.researchgate.net/publication/255485425_Study_on_the_effect_of_chronic_administration_of_Khadirarista_on_different_hematologic_parameters_in_male_Sprague_-Dawley_rats

Chakradhar KV, Datatreya RS. To evaluate the potential of *Yava Paneya Kshara* (Alkali Preparation of Barley) on Nephrolithiasis – A pilot study. *Int Ayur Med J* 2013; 1: 30–6.

Chakrbortty S, Ahmed Z. Study of antimicrobial activity of Ayurvedic and Unani Medicines and their comparative analysis with commercial antibiotics. *Int J Res Appl Nat and Soc Sci* 2013; 1: 63–74.

Chandra D, Mandal AK. Toxicological and pharmacological study of *Navbal Rasayan* – A metal based formulation. *Indian J Pharmacol* 2000; 32: 369–71.

Chandra P, Neetu S, Gangwar AK, Sharma PK. Comparative study of mineralo-herbal drugs (*Kamadugha* and *Sutshekhar Rasa Sada*) on gastric ulcer in experimental rats. *J Pharm Res* 2010; 3: 1659–62.

Chaudhari SY, Ruknuddin G, Biswajyoti JP, Kumar PP. Effect of *tamra bhasma* (calcined copper) on ponderal and biochemical parameters. *Toxicol Int* 2014; 21: 156–9.

Chaudhary PC, Bihari M, Shukla MP. A clinical study on the efficacy of *Vaishvanar Churna* in cases of *Amavata*. *J Res Ayur Siddh* 2002; 23: 36–41.

Chauhan O, Godhwani JL, Khanna NK, Pendse VK. Anti-inflammatory activity of *Muktashukti bhasma*. *Indian J Exp Biol* 1998; 36: 985–9.

Chavan PN. Comparative study of flavonoids from *Chyawanprash* for anti-inflammatory activity. *J Pharm Res* 2011; 4: 1338–9.

Chavan S, Reddy Govind R, Jayant H, Jawale P. A clinical study on efficacy of *Nidigdhikadi kvatha* on *Tamaka Shvasa* (Bronchial asthma). *Int Res J Pharm* 2013; 4: 141–4.

Chitnis KS, Stanley A. Chemical evaluation of *Tamra Bhasma*. *Int J Pharma Bio Sci* 2011; 2: 160–7.

Chittora NK, Shrivastava A, Jain A. Stability-indicating RP-HPLC determination of curcumin in vicco turmeric cream and *Haridrakhand churna*. *Pharmacogn J* 2010; 2: 90–101.

Choudhary AK, Prabhakara Rao G, Nath G, Dixit SK. *Rasa karpura* – an effective antibiotic of Ayurveda. *Sach Ayur* 1999; 51: 769–82.

Chouhan O, Gehlot A, Rathore RK, Choudhary R. Anti-peptic ulcer activity of *Muktashukti Bhasma*. *Pak J Physiol* 2010; 6: 29–31.

Dahikar SB, Bhutada SA, Vibhute SK, Sonvale VC, Tambekar DH, Kasture SB. Evaluation of antibacterial potential of *Trikatu churna* and its ingredients: an *in vitro* study. *Int J Phytomed* 2010; 2: 412–7.

Dama MS, Varshneya C, Dardi MS, Katoch VC. Effect of *trikatu* pre-treatment on the pharmacokinetics of pefloxacin administered orally in mountain Gaddi goats. *J Vet Sci* 2008; 9: 25–9.

Das B, Hazra J. Efficacy of *Nagkesara Churna* and *Godanti bhasma* in non-specific lucorrhoea (*svetapradara*). *Aryavaidyan* 2012; 25: 148–51.

Das S, Shaw BP. Effect of *Dhatri lauha* on gastric acidity of the patients of *amlapitta*. *J Ayur Yoga* 1983; 4: 7.

Das SK, Yadav KD, Dubey SD, Reddy KRC. Anti-microbial study of *Shwasakuthar Rasa*: *In vitro* study. *Int J Ayur Med* 2012; 3.

Dash M, Joshi N, Dwivedi LK, Gupta RS. Spematogenic activity of *Pushpadhanwa Rasa* on Solanum xanthocaropum induced infertility w.s.r. histopathological study. *J Drug Res Ayur Sidd* 2010; 31: 1–18.

Dash M, Joshi N, Dwivedi LK, Gupta RS. Evaluation of *Pushpadhanwa Rasa* a generic Ayurvedic herbomineral formulation in polycystic ovarian syndrome – A pilot study. *J Res Ayur Sidd* 2011; XXXII: 73–98.

Dash M, Joshi N, Gupta RS, Dwivedi LK. Toxicity study of *Naga bhasma* w.s.r. to Ayurvedic measure for toxicity eradication. *Int J Pharm Biol Arch* 2012; 3: 1250–4.

Dash NC, Das BK. Arsenic disulphide (*Rasamanikya*) in the treatment of rheumatoid arthritis (*Amavata*). *J Res Ayu Siddh* 1994; 15: 17–22.

Dash S et al. Clinical trial of *Sutasekhara rasa, Dhatri lauha* and *Kamdudha Rasa* in the management of *Parinamshula*-duodenal ulcer (A review of 109 cases). *J Res Ayurveda Siddha* 1989; 10: 41–9.

Datta GK, Sairam K, Priyambada S, Debnath PK, Goel RK. Antiulcerogenic activity of *Satavari mandur* – an Ayurvedic herbo-mineral preparation. *Indian J Exp Biol* 2002; 40: 1173–7.

Dave KK, De S. Analytical study of *Talisadya Churna*. *J Res Ayur Siddh* 1996; 17: 180–8.

Debabrata C, Shanker K, Pal A, Luqman S, Bawankule DU, Mani D, Darokar MP. Safety evaluation of *Trikatu*, a generic Ayurvedic medicine, Charles Foster rats. *J Toxicol Sci* 2009; 34: 99–108.

Deole YS, Chandola HM. A clinical study on effect of *Brahmi Ghrita* on depression. *AYU* 2008; 29: 207–14.

Deshmukh S, Vyas MK, Harisha CR, Shukla VJ. Pharmacognostical and pharmaceutical analysis of *Shatapushpadya churna*. *Int J Ayur All Sci* 2014; 3: 91–6.

Deshpande PJ, Sharma KR, Singh K. Management of chronic colitis by *Panchamrita parpati kalpa*. *J Res Indian Med* 1977; 12: 1–10.

Devanathan R, Niraimathi KL, Karunanidhi M, Brindha P. Comparative evaluation of *Arka Lavana* – An ayurvedic herbomineral formulation. *Int J Pharm Clin Res* 2013; 5: 37–42.

Devarshi P, Kanase A, Kanase R, Mane S, Patil S, Varute AT. Effect of *mandur bhasma* on lipolytic activities of liver, kidney and adipose tissue of albino rat during CCl4 induced hepatic injury. *J Biosci* 1986; 10: 227–34.

Dhamal S, Wadekar MP, Kulkarni BA, Dhapte VV. Chemical investigations of some commercial samples of calcium based ayurvedic drug of marine origin: *Kapardika Bhasma*. *J Pharm Biol* 2012; 6: 5–12.

Dhirajsingh S Rajput, Rohit A Gokarn, Shukla VJ, Patgiri BJ. P Pharmaceutical standardisation of *Naga Bhasma* prepared by using herbal media. *Ayurpharm Int J Ayur Alli Sci* 2013; 2: 212–23.

Dhundi SN, Patgiri BJ, Prajapati PK. Pharmaceutical standardisation of *Triguna Rasa sindhoora*. 4th World Ayurveda Congress and Arogya Expo proceedings, pp. 88, 9–13 December 2010, Bengaluru, Karnataka, India.

Dhyani PL, Baxi AJ. Physico-chemical and chromatographic Study of *Yavanyrka*. *J Drug Res in Ayur and Sidd* 1985; 6: 2-4: 165–8.

Dighe PM. Efficacy of *Samsharkara Churna* in *Kaphaj Kasa*. *Int J Ayur Med* 2013; 4.

Dixit R, Shivhare GC. Synthetic and analytical studies on *Praval* (Coral: *Tubipora musica*) Bhasam. *BMEBR* 1988; 12: 159–65.

Dixit SK, Jain PC, Joshi D. An experimental study of *Kutajarishta* with special reference to amoebiasis. *Anc Sci Life* 1988; 8: 100–2.

Dixit SK, Nageswara Rao V. Standardization of *Praval bhasma*. *Anc Sci Life* 1998; 17: 1–4.

Dsouza PF, Shenoy A, Moses Samuel R, Shabaraya AR. Anti tumor activity of mercaptopurine in combination with *trikatu* and *gomutra* on 20-Methylcholantrene induced Carcinogenesis. *J Appl Pharm Sci* 2013; 3: 20–4.

Dube CB, Sharma YK. Presence of gold in *Swarna Basant Malti*. *J Res Ayur Siddh* 1980; 2: 274–8.

Dube CB, Kansal CM, Sharma YK. A comprehensive study of *Swarna Basant Malti* in cases of *Rajayakshma* (pulmonary tuberculosis) A clinical and experimental study. *Nagarjun* 1978; 21: 9–14.

Dubey N, Dubey N, Mehta RS, Saluja AK, Jain DK. Antiulcer activity of a traditional pearl preparation: *Mukta Bhasma*. *Res J Pharm Tech* 2009; 2: 287–90.

Dudhamal TS, Gupta SK, Bhuyan C, Singh K. The role of *Apamarga Kshara* in the treatment of *Arsha*. *Ayu* 2010; 31: 232–5.

Dumbre RK, Kale AP, Kamble MB, Patil VR. Effect of *Chandraprabha Vati* in experimental prostatic hyperplasia and inflammation in rats. *J Pharm Res* 2012; 5: 12.

Dushing YA, Shankar L. Antioxidant assessment of *Ashokarishta* – A fermented polyherbal Ayurvedic formulation. *J Pharm Res* 2012; 5: 3165–8.

Dwivedi ML, Tripathi SV, Dwivedi HS. Role of *Phalatrikadi kashaya* and *Arogyavardhini vati* in the management of jaundice (*kamala*). *Sach Ayur* 1984; 37: 87–94.

Dwivedi V, Tripathi BK, Mutsuddi M, Lakhotia SC. Ayurvedic *Amalaki Rasayana* and *Rasa-Sindoor* suppress neurodegeneration in fly models of Huntington's and Alzheimer's diseases. *Curr Sci* 2013; 105: 1711–23.

Elamthuruthy AT, Shah CR, Khan TA, Tatke PA, Gabhe SY. Standardization of marketed *Kumariasava* – an Ayurvedic Aloe vera product. *J Pharm Biomed Anal* 2005; 37: 937–41.

Fulzele SV, Satturwar PM, Joshi SB, Dorle AK. Studies of anti-inflammatory activity of a polyherbal formulation *Jjatyadi ghrita*. *Indian Drugs* 2002; 39: 42–4.

Fulzele SV, Satturwar PM, Joshi SB, Dorle AK. Study of the immunomodulatory activity of *Haridradi ghrita* in rats. *Indian J Pharmacol* 2003; 35: 51–4.
Gahlaut A, Shirolkar A, Hooda V, Dabur R. β-sitosterol in different parts of *Saraca asoca* and herbal drug *Ashokarishta*: Quali-quantitative analysis by liquid chromatography-mass spectrometry. *J Adv Pharm Technol Res* 2013; 4: 146–50.
Gangamma MP Poonam, Rajagopala M. A clinical study on computer vision syndrome and its management with *Triphala* eye drops and *Saptamrita Lauha*. *Ayu* 2010; 31: 236–9.
Gangopadhyay KS, Khan M, Pandit S, Chakrabarti S, Mondal TK, Biswas TK. Pharmacological evaluation and chemical standardization of an ayurvedic formulation for wound healing activity. *Int J Low Extrem Wounds* 2014; 13: 41–9.
Garg LK, Singh M, Singh R, Dhaked R. Pharmaceutico analytical study and comparative antimicrobial effect of *Hartala bhasma* and *Hartalagodanti bhasma*. *Int J Ayur Pharm Res* 2013; 1: 36–46.
Gautam VP, Dubey M, Dass RK, Gupta GP. A clinical study to evaluate the efficacy of *Chandrakala rasa* on *madhumeha*. *Int J Ayur Alli Sci* 2014; 3: 11–6.
Geeta A, Nalini AM. Treatment of *pradara* with *Ashokarishta*. *J Res Ayur Siddh* 1996; 16: 778–80.
Geetha A, Amma A, Nair CPR, Lalitha K. Action of *Pushyanuga Churna* in *pradara*. *J Res Ayur Siddha* 1991; 12: 104–7.
Gehlot A, Mathur AK, Chouhan O. *Dasamula kwatha*: Effect on the analgesic activity of tramadol in rats. *Indian J Pharmacol* 2008; 40: 178–211.
Ghildiyal S, Gautam MK, Joshi VK, Goel RK. Anti-inflammatory activity of two classical formulations of *Laghupanchamula* in rats. *J Ayurveda Integr Med* 2013; 4: 23–7.
Ghildiyal S, Gautam MK, Joshi VK, Goel RK. Analgesic and hypnotic activities of *Laghupanchamula*: A preclinical study. *Ayu* 2014; 35: 79–84.
Ghosal S. Standardization of Ayurvedic drugs and preparations. Proceedings of Captain Srinivasa Murthi Drug Research Institute, Madras, India, 1987; 29–34.
Ghosal S. *Shilajit*: its origin and significance in living matter. *Indian J Indigen Med* 1992; 9: 1–3.
Ghosal S. Shilajit odour, its origin and chemical character. *J Indian Chem Soc* 1994; 71: 360–4.
Ghosal S. The aroma principles of *Gomutra* and *Karpurgandha Shilajit*. *Indian J Indigen Med* 1995; 10: 21.
Ghosal S. Free radical oxidative stress and antioxidative defense. *Phytomedicine* 2000; 21: 1–8.
Ghosal S, Reddy JP, Lal VK. *Shilajit* I: Chemical constituents. *J Pharm Sci* 1976; 65: 772–3.
Ghosal S, Singh SK, Kumar Y, Srivastava RS, Goel RK, Dey R, Bhattacharya SK. Antiulcerogenic activity of fulvic acids and 4"-methoxy-6-carbomethoxybipheny lisolated from shilajit. *Phytother Res* 1988; 2: 187–91.
Ghosal S, Singh SK, Srivastava RS. *Shilajit* part 2. Biphenyl metabolites from *Trifolium repens*. *J Chem Res* 1988b; 196: 165–6.
Ghosal S, Lal J, Singh SK, Dasgupta G, Bhaduri J, Mukopadhyay M, Bhattacharya SK. Mast cell protecting effects of *Shilajit* and its constituents. *Phytotherapy Res* 1989; 3: 249–52.
Ghosal S. Chemistry of *Shilajit*, an immunomodulatory Ayurvedic *rasayan*. *Pure & Appl Chem* 1990; 62: 1285–88.
Ghosal S, Lal J, Singh SK. The core structure of *Shilajit* humus. *Soil Biol Biochem* 1991; 23: 673–80.
Ghosal S, Kawanishi K, Saiki K. The chemistry of *Shilajit* odour. *Indian J Chem* 1995a; 34B: 40–4.
Ghosal S, Lata S, Kumar Y. Free radicals of *Shilajit* humus. *Indian J Chem* 1995b; 34: 591–5.
Goel KN, Singh RH. Clinical trial of *Candanasava* in the treatment of urinary tract infection. *Anc Sci Life* 1991; 10: 248–52.
Goel RK, Banerjee RS, Acharya SB. Antiulcerogenic and anti-inflammatory studies with *Shilajit*. *J Ethnopharmacol* 1990; 29: 95–103.
Gohil H, Dhruve K, Prajapati PK. Role of media in the preparation of *Apamarga Ksharataila*. *AYU* 2010; 31: 391–4.
Gokarn RA, Rajput DS, Yadav P, Galib, Patgiri B, Prajapati PK. Pharmaceutical standardization of *Svarṇa vaṅga*. *Anc Sci Life* 2013; 33: 97–102.
Gokarn RA, Gokarn SR, Hiremath SG. Process standardization and characterization of *Rajata Sindura*. *Ayu* 2014; 35: 63–70.
Gopala Simha KR, Laxminarayana V. Standardization of Ayurvedic polyherbal formulation, *Nyagrodhadi churna*. *Indian J Trad Knowl* 2007; 6: 648–52.
Gosavi DD, Kale R. Effects of *Bilvadi Churna* on Experimentally induced wound healing in rats. *Int J Res Pharm Biomed Sci* 2012; 320–4.

Govindarajan R, Singh DP, Rawat AKS. High-performance liquid chromatographic method for the quantification of phenolics in *Chyavanprash* – A potent Ayurvedic drug. *J Pharm Biomed Anal* 2007; 43: 527–32.

Govindarajan R, Singh DP, Ajay K, Rawat S. Validated RP–LC method for standardization of *Ashokarishta*: A polyherbal formulation. *Chromatographia* 2008; 68: 873–6.

Gunturu KS, Nagarajan P, Strout MP. Ayurvedic herbal medicine and lead poisoning. *J Hematol Oncol* 2011; 4: 51.

Gupta M, Shaw BP, Mukherjee A. A new glycosidic flavonoid from *Jwarhar mahakashay* (antipyretic) Ayurvedic preparation. *Int J Ayurveda Res* 2010; 1: 106–11.

Gupta RA, Singh BN, Singh RN. Pharmacological studies on *Dasamula Kwatha* – Part II. *J Res Ayur Siddh* 1984; 5: 38–50.

Gupta SK, Baghel MS, Bhuyan C, Ravishankar B, Ashok BK, Patil PD. Evaluation of anti-urolithiatic activity of *Pashanabhedadi Ghrita* against experimentally induced renal calculi in rats. *Ayu* 2012; 33: 429–34.

Gupta SS. Effect of *Shilajit*, *Ficus bengalensis* and anti-pituitary extract on glucose tolerance in rats. *Indian J Med Res* 1966; 54: 354–66.

Gupta V, Jain UK. Status of piperine content in Ayurvedic formulation: Method standardization by HPTLC. *Res J Pharm Biol Chem Sci* 2011; 2: 524–30.

Guptai SJ, Kureel MK. Role of *Varunshigru Ghan Vati* in the management of Urinary tract infection (Mutrakrichchra). *J Res Ayur Siddh* 2010; 31: 53–64.

Guruprasad KP, Mascarenhas R, Gopinath PM, Satyamoorthy K. Studies on *Brahma rasayana* in male swiss albino mice: Chromosomal aberrations and sperm abnormalities. *J Ayurveda Integr Med* 2010; 1: 40–4.

Gyawali S, Khan GM, Lamsal R. Evaluation of anti-secretory and anti-ulcerogenic activities of *Avipattikar Churna* on the peptic ulcers in experimental rats. *J Clin Diagn Res* 2013; 7: 1135–39.

Hemadri K. Discovery of *Gomutra* from South India. *Anc Sci Life* 1987; 7: 104.

Hepsibah PT, Rosamma MP, Prasad NB, Kumar PS. Standardisation of the ayurvedic medicine *pinda taila*. *Anc Sci Life* 1996; 15: 222–5.

Hepsibah PT, Prasad NB, Kumar PS. Standardisation of Ayurvedic oils. *Anc Sci Life* 1998; 17: 280–3.

Himmat S, Mishra SK, Pande M. Standardization of *Arjunarishta* formulation by TLC method. *Int J Pharm Sci Rev Res* 2010; 2: 25–8.

Hiremath R, Jha CB, Narang KK. *Vanga Bhasma* and its XRD analysis. *Anc Sci Life* 2010; 29: 24–8.

Hiremath Shobha G, Joshi D, Ray AB. TLC as a tool for standardisation of Ayurvedic formulations with special reference to *Kutajarishta*. *Anc Sci Life* 1993; 12: 358–62.

Honwad Sudheendra V. Toxicity of *Tamra Bhasma* prepared with methods. *Int J Res Ayur Pharm* 2011; 2: 1685–91.

Indu S, Ali M, Onkar S. Chemical Screening, Standardization and Anti-inflammatory Activity of *Shilajit*. Traditional Systems of Medicine, Proceedings of National Workshop on Institute – Industry Interaction on Research in Unani Medicine to Identify Areas of Collaboration, Faculty of Science and Faculty of Medicine (Unani), Jamia Hamdard, Hamdard Nagar, New Delhi, pp. 418–422, 2006. Narosa Publishing House, 22 Daryaganj, New Delhi.

Ingle NM, Ojha NK, Kumar A. Clinical study to evaluate the *Brinhaniya* effect of *Vidarikandadi Yog* to enhance the sport performance in children. *J Ayurveda Integr Med* 2013; 4: 171–5.

Jadar PG, Jagadeesh MS. Preparation and physic-chemical evaluation of *Kshiramandura*. *Anc Sci Life* 2010; 29: 7–12.

Jagtap CY, Ashok BK, Patgiri BJ, Prajapati PK, Ravishankar B. Acute and subchronic toxicity study of *Tamra Bhasma* (Incinerated Copper) prepared from *Ashodhita* (Unpurified) and *Shodhita* (purified) *Tamra* in Rats. *Indian J Pharm Sci* 2013a; 75: 346–52.

Jagtap CY, Ashok BK, Patgiri BJ, Prajapati PK, Ravishankar B. Comparative anti-hyperlipidemic activity of *Tamra Bhasma* (incinerated copper) prepared from *Shodhita* (purified) and *Ashodhita Tamra* (raw copper). *Indian J Nat Prod Res* 2013b; 4: 205–11.

Jagtap CY, Prajapati PK, Patil R, Chaudhary SY. Therapeutic uses of *Tamra* (copper) *bhasma* – A review through *Bhaishajya ratnavali*. *Ayur pharm Int J Ayur Alli Sci* 2014; 3: 128–35.

Jain N, Mishra RN. Antioxidant activity of *Trikatu* mega Ext. *Int J Res in Pharm Biomed Sci* 2009; 2: 624–8.

Jain N, Mishra RN. Adaptogenic activity of *Trikatu* mega Ext. *Int J Res in Pharm Biomed Sci* 2011; 2: 570–4.

Jain P, Rao SP, Singh V, Pandey R, Shukla SS. Acute and sub-acute toxicity studies of an ancient Ayurvedic formulation: *Agnimukha churna*. *Columbia J Pharm Sci* 2014; 1: 18–22.

Jaiswal AK, Bhattacharya SK. Effects of Shilajit on memory, anxiety and brain monoamines in rats. *Indian J Pharmacol* 1992; 24: 12–17.

Janakiraman K, Manavalan R. Studies on effect of coadministration of Trikatu and its components on oral bioavailability of Ampicillin and Norfloxacin in rabbits. *J Pharm Res* 2009; 2: 27–30.

Jayaram Kumar, K, Mouli Nandi. Energy dispersive X-ray spectroscopy in quality control of powdered herbal formulation – *Avipattikar churna*. *J Res Edu Indian Med* 2012; 18: 7–11.

Jha D, Pandey VN. A clinical study on *Kitibh* (Psoriasis). *J Res Ayur Siddh* 1985; 6: 195–212.

Jhanwar B, Solanki R, Singh M. Determination of quality standards for herbal formulation: *Chaturjat churna*. *J Pharmacog Phytochem* 2013; 2: 101–5.

Johri RK, Zutschi U. An Ayurvedic formulation '*trikatu*' and its constituents. *J Ethnopharmacol* 1992; 37: 85–91.

Joshi D, Prabhakara Rao G. Pharmaceutical standardisation of *Rasa karpura* (a non-sulphur mercurial compound). *Sach Ayur* 1992; 45: 214–9.

Joshi GC, Tiwari KC, Pande NK, Pandey G. Bryophytes the source of the origin of *Shilajatu* – A new hypothesis. *Bull Medico-Ethno-Bot Res* 1994; 15: 106–11.

Joshi H, Parle M. *Brahmi rasayana* improves learning and memory in mice. *Evid Based Complement Alternat Med* 2006; 3: 79–85.

Joshi R, Choudhary AK, Prajapati PK, Ravishanka B. Safety study of *Yashada Bhasma*. *J Ayur* 2008; 2: 18–26.

Joukar S, Najafipour H, Dabiri S, Sheibani M, Sharokhi N. Cardioprotective effect of mumie (*shilajit*) on experimentally induced myocardial injury. *Cardiovasc Toxicol* 2014; 14: 214–21.

Jyoti G, Chandola H, Harisha CR, Kalyani R, Shukla VJ. Analytical profile of *Brahmi ghrita*: a polyherbal Ayurvedic formulation. *AYU* 2012a; 33: 289–93.

Jyoti G, Chandola HM, Kalyani R, Shukla VJ. A study to evaluate bacoside A in *Brahmi Ghrita* by HPTLC method. *Int J Green Pharm* 2012b; 6: 184–6.

Kadam A, Maheshwar T, Jha CB. Study of effect of *Abhraka satva bhasma* in alloxan induced hyperglycemia. *Sach Ayur* 2003; 56: 48–53.

Kadam PV, Yadav KN, Patel AN, Navsare VS, Narappanawar NS, Patil MJ. Comparative account of traditionally fermented biomedicine from Ayurveda *Mustakarishta*. *Int J Res Ayur Pharm* 2012a; 3: 429–32.

Kadam PV, Yadav KN, Shivatare RS, Pande AS, Patel AN, Patil MJ. Standardisation of *Gomutra Haritaki vati*: An Ayurvedic formulation. *Int J Pharm Bio Sci* 2012b; 3: 181–7.

Kajaria DK, Gangwar M, Sharma AK, Tripathi YB, Tripathi JS, Tiwari S. Evaluation of *in vitro* antioxidant capacity and reducing potential of polyherbal drug – *Bharangyadi*. *Anc Sci Life* 2012a; 32: 24–8.

Kajaria DK, Tripathi JS, Tiwari SK, Pandey BL. Anti-histaminic, mast cell stabilizing and bronchodilator effect of hydroalcoholic extract of polyherbal compound – *Bharangyadi*. *Anc Sci Life* 2012b; 31: 95–100.

Kalaiselvan V, Shah AK, Patel FB, Shah CN, Kalaivani M, Rajasekaran A. Quality assessment of different marketed brands of *Dasamoolaristam*, an Ayurvedic formulation. *Int J Ayurveda Res* 2010; 1: 10–3.

Kamble S, Dwivedi RR. *In vitro* antibacterial activity of *Chitrakadi Vati* – A herbo-mineral Ayurvedic formulaton against *E. coli*. *Ayur pharm-Int J Ayur All Sci* 2013; 2: 357–63.

Kanase A, Patil S, Thorat B. Curative effects of *Mandura bhasma* on liver and kidney of albino rats after induction of acute hepatitis by CCl4. *Indian J Exp Biol* 1997; 35: 754–64.

Kanikkannan N, Ramarao P, Ghosal S. Shilajit-induced potentiation of the hypoglycaemic action of insulin and inhibition of streptozotocin induced diabetes in rat. *Phytother Res* 1995; 9: 478–81.

Kanojia A, Sharma A, Urimindi V, Gotecha VK. Toxicological evaluation of *Rasa-Sindoor* in albino rats. *Int Ayur Med J* 2013; 1: 1–:5.

Kansal CM, Dube CB. *Samira pannaga rasa* (S.P.R.) in eosinophilia (An experimental study). *J Res Ayur Siddha* 1980; 1: 479–88.

Karan RS, Bhargava VK, Garg SK. Effect of *trikatu* (piperine) on the pharmacokinetic profile of isoniazid in rabbits. *Indian J Pharmacol* 1998; 30: 254–6.

Karan RS, Bhargava VK, Garg SK. Effect of *Trikatu* on the pharmacokinetic profile of indomethacin. *Indian J Pharmacol* 1999a; 31: 160–1.

Karan RS, Bhargava VK, Garg SK. Effect of *Trikatu*, an Ayurvedic prescription, on the pharmacokinetic profile of rifampicin in rabbits. *J Ethnopharmacol* 1999b; 64: 259–64.

Karan RS, Bhargava VK, Garg SK. Effect of *Trikatu*, an Ayurvedic prescription, on the pharmacokinetic profile of carbamazepine in rabbits. *Indian J Physiol Pharmacol* 1999c; 43: 133–6.

Kasture VS, Gharate MK. Chemical and biological evaluation of *Punarnavasava*, A polyherbal Ayurvedic formulation. *World J Pharm Pharmaceut Sci* 2013; 2: 5778–89.

Kataria M, Singh LN. Hepatoprotective effect of Liv.52 and *Kumaryasava* on carbon tetrachloride induced hepatic damage in rats. *Indian J Exp Biol* 1997; 35: 655–7.

Kaushal A, Roy RK, Gattani A, Rao KS, Venkateshwarlu U, Ambulkar PY. Bioavailability of Pravala as calcium supplement in health growing dogs. Presented at 4th World Ayurveda Congress and Arogya, held on 9–13 December 2010, Bengaluru, Karnataka, India.

Keen RW, Deacon AC, Delves HT, Moreton JA, Frost PG. Indian herbal remedies for diabetes as a cause of lead poisoning. *Postgrad Med J* 1994; 70: 113–4.

Keshri A, Verma PR, Prasad CM. Further studies on chemical evaluation of *lauha bhasma* III. *Anc Sci Life* 1996; 16: 26–33.

Khagram R, Mehta CS, Shukla VD, Dave AR. Clinical effect of *Matra Basti* and *Vatari Guggulu* in the management of *Amavata* (rheumatoid arthritis). *Ayu* 2010; 31: 343–50.

Khaksari M, Mahmmodi R, Shahrokhi N, Shabani M, Joukar S, Aqapour M. The effects of Shilajit on brain edema, intracranial pressure and neurologic outcomes following the traumatic brain injury in rat. *Iran J Basic Med Sci* 2013; 16: 858–64.

Khanvilkar V, Patil L, Kadam V. Standardization of *Chitrakadi Vati*: An Ayurvedic polyherbal formulation. *Int J Pharm Sci Drug Res* 2014; 6: 303–9.

Khot BM, Patil AJ, Kakad AC. Comparative clinical study of *Dhatri Lauha* and *Navayasa Lauha* in *Garbhini Panduroga* with reference to anemia in pregnancy. *IOSR J Dental Med Sci* 2013; 11: 28–33.

Kong YC, Butt PPH, Ng KH, Cheng KF, Camble RC, Malla SB. Chemical studies on a Napalese panacea; Shilajit. *Int J Crude Drug Res* 1987; 25: 179–87.

Kotabagi JD, Vaidya SS, Santhosh B, Jadar PG. Screening of hepatoprotective activity of *Lauha Parparti*. *Aryavaidyan* 2010; 23: 157–9.

Kotabagi JD, Vaidya SS, Santhosh B, Jadar PG. Comparative study of hepatoprotective activity of Lauha parpati prepared by four different methods. *Int J Res Ayur Pharm* 2011; 2: 354–57.

Kotecha KN, Kotecha BK, Shukla VJ, Prajapati PK, Ravishankar B. Acute toxicity study of *Vasaguduchyadi Kwatha*: A compound Ayurvedic formulation. *Ayu* 2013; 34: 327–30.

Kotrannavar V, Sarashetty R, Kanthi V. Physico-chemical analysis of *Mayurapuccha Bhasma* prepared by two methods. *Anc Sci Life* 2012; 32: 45–8.

Kumar A, Kumar N. To evaluate effect of *Palasha Kshara* in the management of urolithiasis. *J Res Ayur Siddh* 1995; 16: 43–50.

Kumar A, Kumar N. Effects of a herbo-mineral combination (*sunthi guggulu godanti*) in amavata (rheumatoid arthritis). *J Res Ayur Siddh* 2003; 24: 14–30.

Kumar A, Garai AK. A clinical study on Pandu Roga, iron deficiency anemia, with *Trikatrayadi Lauha* suspension in children. *J Ayurveda Integr Med* 2012; 3: 215–22.

Kumar G, Srivastava A, Sharma SK, Gupta YK. Safety and efficacy evaluation of Ayurvedic treatment (*Arjuna* powder and *Arogyavardhini Vati*) in dyslipidemia patients: A pilot prospective cohort clinical study. *Ayu* 2012a; 33: 197–201.

Kumar G, Srivastava A, Sharma SK, Gupta YK. Safety evaluation of an Ayurvedic medicine, *Arogyavardhini vati* on brain, liver and kidney in rats. *J Ethnopharmacol* 2012b; 140: 151–60.

Kumar G, Srivastava A, Sharma SK, Gupta YK. The hypolipidemic activity of Ayurvedic medicine, *Arogyavardhini vati* in Triton WR-1339-induced hyperlipidemic rats: A comparison with fenofibrate. *J Ayurveda Integr Med* 2013; 4: 165–70.

Kumar G, Srivastava A, Sharma SK, Gupta YK. Safety evaluation of mercury based Ayurvedic formulation (*Sidh Makardhwaj*) on brain cerebrum, liver & kidney in rats. *Indian J Med Res* 2014; 139: 610–8.

Kumar KG, Parmar G, Patgiri BJ. Pharmaceutical standardization of *Jala Shukti Bhasma* and *Mukta Shukti Bhasma*. *Ayu* 2012; 33: 136–42.

Kumar KV, Kumar PS, Ahammed Basheer PK, Safvan K, Paredath S, Chaliyath S, Manivannan R, Lalitha KG. Comparative antidiabetic investigation of Ayurveda polyherbal formulation, *Nisamalaki Churna* tablet in alloxan-induced diabetic rats. *Int J Biol Pharm Res* 2012; 3: 586–92.

Kumar S, Singh G, Reddy KR. Effect of *Drakshavaleha* in cyclophosphamide induced weight loss and reduction in crown-rump length in developing mice embryo. *Ayu* 2013; 34: 215–9.

Kumar SVK, Mishra MS. Hepatoprotective activity of the *trikatu churna*: an ayurvedic formulation. *Indian J Pharm Sci* 2004; 66: 365–7.

Kumar T, Chandrashekar KS, Tripathi DK, Nagori K, Pure, Agrawal S, Ansari TJ. Standardization of "*Gokshuradi Churna*": An ayurvedic polyherbal formulation. *J Chem Pharm Res* 2011; 3: 742–9.

Kumar T, Larokar YK, Jain V. Standardization of different marketed brands of *Ashokarishta*: An Ayurvedic formulation. *J Sci Innov Res* 2013; 2: 993–8.

Kumar Y, Singh BM, Gupta P. Clinical and metabolic markers based study of *Swas Kasa Chintamani Rasa* (An Ayurvedic herbo-metallic preparation) in childhood bronchial asthma (*Tamak Swas*). *Int J Green Pharm* 2014; 8: 37–44.

Ladha KS, Kasar RP, Chaudhary J, Shukla A. A HPTLC densitometric determination of antioxidant constituents from *Chyawanprash*. *Indian Drugs* 2008; 45, 7 (article 3).

Lagad CE, Ingole R. Pharmaceutical and clinical evaluation on *Vanga bhasma* in the management of madhumeha (diabetes mellitus). *AYU* 2009; 30: 443–6.

Lakshmi Narasimha Reddy N, Yamini K, Gopal V. Anthelmintic of aqueous and ethanolic extract of *Trikatu churna*. *J Appl Pharma Sci* 2011; 1: 140–2.

Lal UR, Tripathi SM, Jachak SM, Bhutani KK, Singh IP. HPLC Analysis and standardization of *Arjunarishta* – An Ayurvedic cardioprotective formulation. *Sci Pharm* 2009; 77: 605–16.

Lal UR, Tripathi SM, Jachak SM, Bhutani KK, Singh IP. Chemical changes during fermentation of *Abhayarishta* and its standardization by HPLC-DAD. *Nat Prod Comm* 2010; 5: 575–9.

Lal VK, Panday KK, Kapoor ML. Literary support to the vegetable origin of *Shilajit*. *Anc Sci Life* 1988; 7: 145–8.

Lala LG, D'Mello PM, Naik SR. Pharmacokinetic and pharmacodynamic studies on interaction of "*trikatu*" with diclofenac sodium. *J Ethnopharmacol* 2004; 91: 277–280.

Lavekar GS. Clinical safety and efficacy of *Dhatri Lauha*: a classical Ayurvedic formulation in iron deficiency anaemia, Pandu Roga. Central Council for Research in Ayurveda and Siddha, Department of AYUSH, Ministry of Health & Family Welfare, Government of India, 2010.

Lavekar GS, Ravishankar B, Rao SV, Gaidhani SN, Ashok BK, Shukla VJ. Safety/Toxicity studies of ayurvedic formulation – Navratna rasa. *Toxicol Int* 2009; 16: 37–42.

Lavekar GS, Ravishankar B, Gaidhani S, Shukla VJ, Ashok BK, Padhi MM. *Mahayograj guggulu*: Heavy metal estimation and safety studies. *Int J Ayurveda Res* 2010; 1: 150–8.

Louis T, Yuvraj P, Madhavchandran V, Gopinatha N. Study of subacute toxicity of *Kajjali*, A combination of mercury and sulphur on albino rats. *Int Ayu Med J* 2014; 2: 936–9.

Madaan S, Sharma ML. A comparative study on clinical evaluation of *Anu Taila* and *Kaphaketu rasa* w.s.r. to *Ardhavabhed* (Migraine). *J Res Ayur Siddh* 2005; 26: 66–74.

Mahajan KN, Singhai AK, Vadnere GP. Investigation on anticataract activity of *Triphala Ghrita*. *E-J Chem* 2011; 8: 1438–43.

Mahto RR, Dave AR, Shukla VD. A comparative study of *Rasona Rasnadi Ghanavati* and *Simhanada Guggulu* on *Amavata* with special reference to Rheumatoid arthritis. *Ayu* 2011; 32: 46–54.

Makhija IK, Shreedhara CS, Aswatha Ram HN. Mast cell stabilization potential of *Sitopaladi churna*: An ayurvedic formulation. *Pharmacog Res* 2013; 5: 306–8.

Mallappanavar S. Role of *Nishadi Taila* in the management of *Fistula-in-Ano*. *Ann Ayur Med* 2013; 2: 147–55.

Mamun Al-Amin Md, Choudhuri SK, Das BK, Hannan JMA. Toxicological and pharmacological effect of *Khadirarista* in animal model. http://stmjournals.com/med/index.php?journal=AYUSH&page=article&op=view&path%5B5D=203

Mandavkar YD, Jalalpure SS. Effect of *Amavatavidhvansa rasa*: A herbo-mineral formulation on carrageenan-induced inflammation in rats. *Indian J Health Sci* 2014; 7: 12–4.

Manjunatha S, Jaryal AK, Bijlani RL, Sachdeva U, Gupta SK. Effect of *Chyawanprash* and vitamin C on glucose tolerance and lipoprotein profile. *Indian J Physiol Pharmacol* 2001; 45: 71–9.

Manjusha R, Vaghela DB, Shukla VJ. A clinical study on the role of *Akshi Tarpana* with *Jeevantyadi Ghrita* in *Timira* (Myopia). *Ayu* 2011; 32: 540–5.

Manmode R, Manwar J, Vohra M, Padgilwar S, Bhajipale N. Effect of preparation method on antioxidant activity of ayurvedic formulation *Kumaryasava*. *J Homeo Ayur Med* 2012; 1: 114. doi: 10.4172/2167-1206.1000114.

Mashru M, Galib R, Shukla VJ, Ravishankar B, Prajapati PK. Effect of *Sameera Pannaga Rasa* (arsenomercurial formulation) in the management of *Tamaka Shwasa* (bronchial asthma) – Randomized double blind clinical study. *Ayu* 2013; 34: 346–51.

Maurya DPS, Sannd BN, Swaroop R. Anti-diuretic study of *Nardiyalaxmi Vilas Ras Mishran* in albino-rats. *J Res Ayur Siddh* 1982; 3: 184–6.

Md. Mamun Al-Amin, Syed Zaheed Kamal. Biochemical studies of *Raktapittantaka Louha*. *Int J Ayur Res* 2013; 4.

Meena AK, Rao MM, Panda KP, Yadav A, Singh U, Singh B. Standardisation of Ayurvedic polyherbal formulation, *Pancasama Churna*. *Int J Pharmacog Phytochem Res* 2010; 2: 11–4.

Meena AK, Mangal AK, Rao MM, Panda P, Simha GV, Shakya SK, Padhi MM, Ramesh B. Evaluation of standardization parameters for *Sitopaladi Churna* an Ayurvedic Formulation. *Asian J Res Chem* 2011; 4: 1867–71.

Meena AK, Simha GV, Mangal MK, Sannd R, Panda P, Rao MM, Padhi MM. Evaluation of quality control parameters for *Srngyadi Churna* – A potential ayurvedic formulation. *Res J Pharmcog Phytochem* 2013; 5: 42–6.

Mehra R, Makhija R, Vyas N. A clinical study on the role of *Ksara Vasti* and *Triphala Guggulu* in *Raktarsha* (Bleeding piles). *Ayu* 2011; 32: 192–5.

Mehta N, Patgiri BJ, Ravishankar B, Prajapati PK. The effect of *Rasa karpuradrava* on *kshudra kustha*. *Int J Ayur Med* 2011; 2: 81–92.

Mehta NJ, Ashok BK, Ravishankar B, Prajapati PK. A comparative dermal toxicity evaluation of *Rasa karpura* and mercuric chloride in rats. *Global J Res Med Plants Indigens Med* 2012; 1: 448–56.

Menon A, Krishnan Nair CK. Ayurvedic formulations ameliorate cisplatin-induced nephrotoxicity: preclinical studies on *Brahma Rasayana* and *Chyavanaprash*. *J Cancer Res Ther* 2013; 9: 230–4.

Mishra A, Mishra AK, Tiwari OP, Jha S. HPLC analysis and standardization of *Brahmi vati* – An Ayurvedic poly-herbal formulation. *J Young Pharm* 2013; 5: 77–82.

Mishra A, Mishra AK, Tiwari OP, Jha S. In-house preparation and characterization of an Ayurvedic bhasma: *Praval bhasma*. *J Integr Med* 2014; 12: 52–8.

Mishra DD, Holla BV, Kishore P. Field clinical trial of *Nityananda rasa* in the treatment of *slipada*. *Sach Ayur* 1979; 31: 919–22.

Mishra PK, Rai NP. Effect of castor oil along with *Ajmodadi Churna* & *Ruksha Sweda* in the management of rheumatoid arthritis vis-a-vis *Amavata*. *Int J Pharm Res Scholars* 2014a; 3: 566–72.

Mishra PK, Rai NP. Prognostic effect of *Ajmodadi churna* in the management of rheumatoid arthritis vis-a vis *amavata*. *Univ J Pharm* 2014b; 3: 47–9.

Mishra RP, Mishra S. *In Vitro* assessment of genotoxicity of some ayurvedic drugs in human lymphocytes by using single cell gel electrophoresis. *J Drug Metab Toxicol* 2013; 4: 139.

Mitra A, Chakraborty S, Auddy B, Tripathi P, Sen S, Saha AV, Mukherjee B. Evaluation of chemical constituents and free-radical scavenging activity of *Swarnabhasma* (gold ash), an ayurvedic drug. *J Ethnopharmacol* 2002; 80: 147–53.

Mohaptra S, Jha CB. Physicochemical characterization of Ayurvedic bhasma (*Swarna makshika bhasma*): An approach to standardization. *Int J Ayurveda Res* 2010; 1: 82–6.

Mohapatra S, Jha CB. Evaluation of the effect of conventionally prepared *Swarna makshika bhasma* on different bio-chemical parameters in experimental animals. *J Ayurveda Integr Med* 2011; 2: 187–91.

Mukherjee S, Pawar N, Kulkarni O, Nagarkar B, Thopte S, Bhujbal A, Pawar P. Evaluation of free-radical quenching properties of standard Ayurvedic formulation *Vayasthapana Rasayana*. *BMC Complement Altern Med* 2011; 11: 38.

Mulik SB, Jha CB. Physicochemical characterization of an Iron based Indian traditional medicine: *Mandura Bhasma*. *Anc Sci Life* 2011; 31: 52–7.

Murunikkara V, Rasool M. Trikatu, an herbal compound that suppresses monosodium urate crystal-induced inflammation in rats, an experimental model for acute gouty arthritis. *Cell Biochem Funct* 2014a; 32: 106–14.

Murunikkara V, Rasool MK. Trikatu, an herbal compound mitigates the biochemical and immunological complications in adjuvant-induced arthritic rats. *Int J Rheum Dis* 2014b. doi: 10.1111/1756-185X.12535. [Epub ahead of print]

Muzaffer A, Dasan KKS, Ramar C, Usman Ali S, Purushothaman KK. Experimental studies on the fermentation in asavas and aristas part II – *Drakshasva*. *Anc Sci Life* 1983; 2: 216–9.

Nafisa B, Bhan A, Paradkar NS, Shaikh A, Nandedkar TD, Bhutani KK, Roy-Chaudhury M. Postnatal development and reproductive performance of F1 progeny exposed *in utero* to an ayurvedic contraceptive: *Pippalyadi Yoga*. *J Ethnopharmacol* 2007; 109: 406–11.

Nagarajan S, Pemiah B, Krishnan UM, Rajani KS, Krishnaswamy S, Sethuraman S. Physico-chemical characterization of lead based Indian traditional medicine – *Naga Bhasma*. *Int J Pharm Pharm Sci* 2012; 4: 69–74.

Nagarajan S, Krishnaswamy S, Pemiah B, Rajan KS, Krishnan U, Sethuraman S. Scientific insights in the preparation and characterisation of a lead-based *Naga Bhasma*. *Indian J Pharm Sci* 2014; 76: 38–45.

Nagaraju V, Joshi D, Aryya N. Toxicity studies on *vanga bhasma* (Part I – with special reference to G. I. T. Liver and Pancreas). *Ancient Sci Life* 1984; 5: 32–35.

Nagaraju V, Joshi D, Aryya NC. A study on the *vrysa* property (Testicular regenerative potential) of *vanga bhasma*. *Anc Sci Life* 1985; 5: 42–8.

Nagarsi T. Pharmaceutical study on naga (lead metal) and its role in the management of diabetes mellitus. PhD thesis submitted to Banaras Hindu University, 1992.

Nageswar V, Dixit SK. Pharmaceutical standardisation of *Godanti bhasma*. *Sach Ayur* 1996; 48: 1101–3.

Naphade A, Bhuyan G, Murthy PSC. A clinical study to assess the efficacy of *Malla Sindur* for the management of *Pakshaghata* (Hemiplegia). *J Res Ayur Sidd* 2011; 32: 67–78.

Nariya MB, Parmar P, Shukla VJ, Ravishankar B. Toxicological study of *Balacaturbhadrika churna*. *J Ayurveda Integr Med* 2011; 2: 79–84.

Nazar M. Antidiabetic activity of hydroalcoholic extract of herbal marketed product *Madhuhari Churna* in alloxan-induced diabetic rats. *Int J Pharm Biol Arch* 2010; 1: .

Neetu D, Singhal H, Mitra S, Joshi N, Sharma KC. Accelerated stability study of *Gojihvadi Kwath* granules. *Int J Pharm Biol Arch* 2011; 2: 1224–9.

Niranjan K, Ravishankara MN, Padh H, Rajani M. A new spectrophotometric method for the estimation of total alkaloids in the stem bark and seed of *Holarrhena antidysenterica* (Linn.) Wall. and in the Ayurvedic formulation kutajarishta. *J Nat Rem* 2002; 2: 2.

Nishat N, Muhammad S, Abdullah Md. SH, Chowdhury SK. Evaluation of CNS effects of *Dhatri Lauha*: An Ayurvedic preparation. *Int J Drug Dev Res* 2012; 4: 234–40.

Ojha JK, Bajpai HS, Sharma PV, Khanna MN, Shukla PK, Sharma TN. *Chyavanprash* as an anabolic agent – experimental study (preliminary work). *J Res Indian Med* 1973; 8: 11–4.

Ojha JK, Sharma PV, Bajpai HS. *Lavangadi vati* as an antitussive agent. *J Res Indian Med* 1979; 14: 25–32.

Onkar P, Bangar J, Karodi R. Evaluation of antioxidant activity of traditional formulation *Giloy satva* and hydroalcoholic extract of the *Curculigo orchioides* Gaertn. *J Appl Pharm Sci* 2012; 2: 209–13.

Otta SP, Tripaty RN. Clinical trial of *Phalaghrita* on female infertility. *Anc Sci Life* 2012; 22: 56–63.

Pallavi S, Jamadagni PS, Jamadagni SB, Hazra J. Repeated dose oral toxicity of *Trivanga Bhasma* in Swiss albino mice. *AYU* 2013; 34: 118–23.

Panda AK, Doddanagali SR. Clinical efficacy of herbal *Padmapatradi yoga* in bronchial asthma (*Tamaka Swasa*). *J Ayurveda Integr Med* 2011; 2: 85–90.

Panda SK, Das S, Behera M, Tripathi B, Patil D. Standardization of *Sitopaladi Churna*: A poly-herbal formulation. *Der Pharmacia Lettre* 2012; 4: 205–16.

Pandey HC, Tewari LC. Latex of *Euphorbia royleana* Boiss., the source of *Gomutra Shilajit* (*Shilajatu*), an ancient miraculous drug of India. *Quart J Crude Drug Res* 1975; 13: 135–42.

Pandey MP. Comparative evaluation of *Hinguleshwar Rasa* and *Tribhuvan Kirti Rasa* in 'Flu-syndrome. *Nagarjun* 1968; 1: 35–8.

Pandey NN. Standardisation of *Swarna parpati*, an Ayurvedic gold preparation. *Sach Ayur* 1982; 35: 45–7.

Pandey NN, Shukla SS, Sharma U. Chemistry of *Rasamanikya* – a preliminary study. Aryavaidyan 1999; 13: 30–4.

Pandey SA, Joshi NP, Pandya DM. Clinical efficacy of *Shiva Guggulu* and *Simhanada Guggulu* in Amavata (Rheumatoid Arthritis). *Ayu* 2012; 33: 247–54.

Pandit S, Suri TK, Jana U, Bhattacharyya D, Debnath PK. Anti-ulcer effect of *shankha bhasma* in rats: a preliminary study. *Indian J Pharmacol* 2000; 32: 378–80.

Pandya S, Dhiman K. Role of *Pippalyadi Yoga* in infertility caused by anovulatory factor. *J Homeo Ayur Med* 2014; doi: 10.4172/2167-1206.S1-001.

Pandya HK, Dhiman K, Dei LP, Thakar AB, Harisha CR. Pharmacognostical and phyto-chemical evaluation of *Pippalyadi Yoga*: A polyherbal formulation. *Int J Ayur Pharma Res* 2014; 2: 116–23.

Panigrahi H, Sharma SS, Kushwaha HK. Clinical study on *Rohitakarista* in reference to cholelithiasis. *J Res Ayur Siddha* 2004; 25: 44–52.

Panigrahi, HK. Efficacy of Ayurvedic medicine in the treatment of uncomplicated chronic sinusitis. *Anc Sci Life* 2006; 26: 6–11.

Parameswaran S, Mandar ND. Standardisation of an Ayurvedic formulation "*Sanjivani Vati*". *Int J Green Pharm* 2010; 4: 153–5.

Paranjpe P, Kulkarni PH. Comparative efficacy of four Ayurvedic formulations in the treatment of acne vulgaris: a double-blind randomised placebo-controlled clinical evaluation. *J Ethnopharmacol* 1995; 127–32.

Parihar SS, Mishra SK, Singh H. Standardisation of *Ashokarishta* formulation by TLC method. *Int J Pharm Tech Res* 2010; 2: 1427–30.

Parimi S, Dixit SK, Gode KD, Joshi D. Anti-diabetic effect of *chandraprabha vati* – a reappraisal (experimental study). *Sachitra Ayur* 1995; 48: 395–9.

Park JS, Kim GY, Han K. The spermatogenic and ovogenic effects of chronically administered *Shilajit* to rats. *J Ethnopharmacol* 2006; 107: 349–53.

Parle M, Bansal N. Antiamnesic activity of an Ayurvedic formulation *Chyawanprash* in mice. *Evid Based Complement Alternat Med* 2011; 2011: 898593.

Parmar G, Parmar M, Ramesh AC. Comparative clinical study of *Palasha Kshara* Sutra and *Apamarga Kshar Sutra* in the management of fistula-in-ano. *Int J Res Ayur Pharm* 2014; 5: 280–3.

Patankar U, Patankar U, Sharma SS. Role of *jalaukavacharan* and *Pancha Nimba Churna Vati* in *vicharchika*. *Sach Ayur* 2005b; 58: 364–7.

Patel BR, Ashok BK, Ravishankar B. Study on the diuretic activity of *Veerataru Kwatha* in albino rats. *Ayu* 2011; 32: 395–7.

Patel VR, Patel RK. Simultaneous analysis and quantification of markers of *Manjisthadi churna* using high performance thin layer chromatography. *Indian J Pharm Sci* 2013; 75: 106–9.

Patgiri B, Dutta SK, Sahai M, Jha CB. Study of *Arogyavardhini vati* with special reference to its analytical study. *Sachitra Ayur* 1999; 51: 855–9.

Patgiri B, Umretia BL, Vaishnav PU, Prajapati PK, Shukla VJ, Ravishankar B. Anti-inflammatory activity of Guduchi Ghana (aqueous extract of *Tinospora cordifolia* Miers.). *Ayu* 2014; 35: 108–10.

Patigiri V, Aryya NC, Jha CB. Study of *Arogyavardhini vati* with special reference to its toxicity study. *Sachitra Ayur* 2001; 53: 694–6.

Patil NB, Hiremath SK, Ekbote NN, Lingayat A. Physicochemical analysis of *Dashanga Agada* – An Ayurvedic formulation. *IOSR J Pharm* 2013; 3: 20–5.

Patra KC, Pareta SK, Harwansh R, Kumar M, Meena KP. Evaluation of *Shivaksharpachan Churna* for its gastroprotective activity. *Pharmacologyonline* 2011; 2: 731–37.

Pattan Shetty JK, Pushpalatha H, Bikshapathi T. Study of *Arogyavardhini* and *Gandhaka rasayana* in the treatment of leucoderma. *Sach Ayur* 2000; 53: 438–44.

Pattan Shetty JK, Pushpalatha H, Bikshapathi T. Study of *Arogyavardhini* and *Gandhaka rasayana* in the treatment of leucoderma. *Sach Ayur* 2000; 53: 438–44.

Pattanaik N, Singh AV, Pandey RS, Singh BK, Kumar M, Dixit SK, Tripathi YB. Toxicology and free radicals scavenging property of *Tamra Bhasma*. *Indian J Clin Biochem* 2003; 18: 181–9.

Pattanayak P, Panda SM, Dash S, Behera M, Mishra SK. Study of antitussive activity of *Sitopaladi churna*: A polyherbal formulation. *Int J Pharma Sci Rev Res* 2010; 4: 65–7.

Pattonder, Ranjan K, Chandola HM, Vyas SN. Clinical efficacy of *Shilajatu* (asphaltum) processed with *Agnimantha* (*Clerodendrum phlomidis* Linn.) in *sthaulya* (obesity). *AYU* 2011; 32: 526–31.

Paul W, Sharma CP. Blood compatibility studies of *Swarna bhasma* (gold bhasma), an Ayurvedic drug. *Int J Ayur Res* 2011; 2: 14–22.

Prajapathi ML, Sharma AK, Sastry CHS. Clinical evaluation of *Palasha Pushpadi Churna* in the management of *Madhumeha Roga* (Diabetes mellitus). *J Res Ayur Siddh* 2000; 21: 1–10.

Prakash B, Parikh PM, Pal SK. Herbo-mineral ayurvedic treatment in a high risk acute promyelocytic leukemia patient with second relapse: 12 years follow up. *J Ayurveda Integr Med* 2010; 1: 215–8.

Prakash B, Parikh PM, Pal SK. Herbo-mineral ayurvedic treatment in a high risk acute promyelocytic leukemia patient with second relapse: 12 years follow up. *J Ayurveda Integr Med* 2010; 1: 215–8.

Prameela Devi K. Clinical evaluation of *Pushyanuga churna* & *Lodhrasava* in *rakta pradara* (DUB). *Indian J Trad Knowl* 2007; 6: 429–31.

Prameela Devi K. Clinical evaluation of *Pushyanuga churna* & *Lodhrasava* in *rakta pradara* (DUB). *Indian J Trad Knowl* 2007; 6: 429–31.

Prasad GP, Amaranath, Babu G, Lakshmi VN, Maheswa T, Sai Prasad AJV, Swamy GK. Effect of external application of *Ashthamulika Taila* in the symptomatic treatment of *shleepada* (filriasis). *J Res Ayur Siddh* 2011; 32: 115–30.

Prasad VVRD, Rao PG, Joshi D. Chemical study of *Rasa karpura* (a mercurial preparation). *Sach Ayur* 1992; 44: 591–3.

Prasad Y, Murthy PHC, Kumar BD. Pharmaceutical preparation and toxicological study of *Rasa Bhasma*. *Int J Ayur Pharma Res* 2013; 1: 36–46.

Prasanna Kumar T, Vijay Kumar GS, Yumnam Dhanesori Devi. *In-vitro* antifungal activity of *Gandhaka Rasayana*. *Int J Ayur* 2010; 1.

Prasanna Kumari P, Rama Sastry VVS, Vijaya Babu V, Ravindar K. Effect of *Kayyonyadi churna* in the management of *Pandu roga* (Anaemia). *Int J Ayur Res* 2013; 4.

Pravin MT, Patgir BJ, Prajapati PK. Pharmaceutical standardization of *Naga Bhasma*. *AYU* 2009; 30: 300–309.

Raghuveer V, Rao N. Antimicrobial activity of *Sheetamshu Rasa*. *IAMJ* 2014; 2: 885–7.

Ragvani H, Bhatt N, Thakar A. Clinical evaluation of *Vyoshadi Guggulu* and *Haritaki Churna* in the management of dyslipidemia. *Int J Pharm Biol Arch* 2013; 4: 643–6.

Rai PD, Rajput SJ. Biological evaluation of polyherbal ayurvedic cardiotonic preparation "*Mahamrutyunjaya rasa*". *Evid Based Complement Alternat Med* 2011; 2011. pii: 801940. doi: 10.1155/2011/801940. Epub 2010 Sep 2.

Rajagopalan K. *Shilajita* (asphaltum). *Anc Sci Life* 1984; 4: 1 (Editorial).

Rajagopala S, Ashok BK, Ravishankar B. Immunomodulatory activity of *Vachadhatryadi Avaleha* in albino rats. *Ayu* 2011; 32: 275–8.

Rajalakshmi D, Brindha P. Chemical standardisation studies on *varatika bhasma*. *Int J Curr Pharm Res* 2010; 2: 12–6.

Rajalakshmy MR, Sindhu A. Preliminary screening and antioxidant activity of an Ayurvedic formulation: *Balarishtam*. *Int J Res Ayur Pharm* 2011; 2: 1645–7.

Rajani J, Ashok BK, Galib, Patgiri BJ, Prajapati PK, Ravishankar B. Immunomodulatory activity of *Amalaki Rasayana*: An experimental evaluation. *Anc Sci Life* 2012; 32: 93–8.

Rajasekaran A, Murugesan S. Standardization of *Shringa Bhasam*. *Anc Sci Life* 2002; 21: 167–9.

Rajnath, Prasad B. A note on the occurrence of *Shilajit*. Indian Science Congress, 1942.

Rajoria K, Singh SK, Sharma RS, Sharma SN. Clinical study on *Laksha Guggulu*, Snehana, Swedana & Traction in Osteoarthritis (Knee joint). *Ayu* 2010; 31: 80–7.

Rajput D, Patgiri BJ, Galib R, Prajapati PK. Anti-diabetic formulations of *Naga bhasma* (lead calx): A brief review. *Anc Sci Life* 2013; 33: 52–9.

Ramachandran AP, Prasad SM, Samarakoon SM, Chandola HM, Harisha CR, Shukla VJ. Pharmacognostical and phytochemical evaluation of *Vara Asanadi Kwatha*. *Ayu* 2012; 33: 130–5.

Ramnath V, Rekha PS. *Brahma Rasayana* enhances *in vivo* antioxidant status in cold-stressed chickens (Gallus gallus domesticus). *Indian J Pharmacol* 2009; 41: 115–9.

Randhir, Sharma LN. Clinical study of early ejaculation and its management with *Virya Sthambhak Vati*. *J Res Ayur Siddh* 2010; 31: 23–34.

Rao KS, Vaish P. Standardization of *Rasa parpati*. *Int J Appl Ayur Res* 2014; 1: 1–6.

Rao MM, Kar AC, Bhattacharya P, Hazra J. Comparative therapeutic evaluation of the efficacy of *pippali vardhaman ksheerapaka* + *sameerapannaga* rasa vs. *shunti* + *guggulu* + *godanti* in the management of *Amavata* (rheumatoid arthritis). *J Res Ayur Siddh* 2006; 27: 81–94.

Rao N, Shah J, Patel K, Gandhi TE. Evaluation of the efficacy of *Hajrul Yahud Bhasma* in urolithiasis using rat as an experimental model. *Ind J Pharmacol* 2011; 40: 178–211.

Rao P, Amrit I, Jain P, Singh V. Antiulcer activity of *Agnimukha churna*. *Int J Ayur Pharma Research* 2014; 2: 40–6.

Rao VN, Suresh P, Dixit SK, Gode KD. Effect of *yashada bhasma* in streptozotocin-induced diabetes. *Anc Sci Life* 1997; 17: 114–6.

Rasheed A, Satyanarayana KV, Gulabi PS, Rao MS. Chemical and pharmacological standardization of *Ashwagandhadi lehyam*: an ayurvedic formulation. *J Complement Integr Med* 2013; 15: 10.

Rasheed A, Naik M, Mohammed-Haneefa KP, Arun-Kumar RP, Azeem AK. Formulation, characterization and comparative evaluation of *Trivanga bhasma*: a herbo-mineral Indian traditional medicine. *Pak J Pharm Sci* 2014a; 27: 793–800.

Rasheed A, Sri MT, Mohammed-Haneefa KP, Arun-Kumar RP, Azeem AK. Formulation, standardization and pharmacological studies of *Saraswataristam*: A polyherbal preparation. *Pak J Pharm Sci* 2014b; 27: 1163–9.

Rathi AR, Khatoon, Rawat S, Singh AK, Mehrotra S, Pandey MM. Evaluation of Ayurvedic compound formulations III – *Laghugangadhar Churna*. *Indian J Trad Knowl* 2010; 9: 576–80.

Ratnasooriya WD, Weerasekera KR, Dhammarathana I, Madampe H, Tissera AK, Hettiarachchige Ariyawansha S. Diuretic activity of chandraprabha vati an Ayurvedic herbo-mineral formulation in Rats. *World J Pharm Sci* 1992; 2014; 2.

Ravishankar AG, Mahesh TS. *Tankana bhasma kavala* in chronic tonsillitis. *Unique J Ayur Herbal Med* 2013; 01: 41–4.

Ravishankar B, Ravi M. Evaluation of *Makardhwaj Rasayan* Tablet for safety profile. http://www.virgouap.com/pdffile/Toxicity_Reports_Makardhwaj_Rasayan_Tablet.pdf

Ravishankar B, Sasikala CK. Pharmacological studies on *Lodhrasava*. *J Res Ayu Siddh* 1986; 7: 33–46.

Ravishankar B, Sridhar BN, Vijay kumar D. Evaluation of compound Ayurvedic preparations for anti-pyretic, analgesic and anti-inflammatory effect. *J Res Ayur Siddh* 1986; 7: 136–145.

Ravishanker B. Pharmacological evaluation of compound Ayurvedic preparations Part-B – *Mandura Vataka*. *J Res Ayur Siddh* 1984; 5: 51–62.

Rawal JH. Clinical study of *Pippalyadi Yoga* as contraceptive method. *Sach Ayur* 1991; 44: 353–9.

Rayees S, Johri RK, Singh G, Tikoo MK, Singh S, Sharma SC, Satti NK, Gupta VK, Bedi YS. Sitopaladi, a poly-herbal formulation inhibits IgE mediated allergic reactions. *Spatula DD* 2012; 2: 75–82.

Reddy BU, Seetharam YN. Anthelmintic activity of *Trikatu Churna* and its ingredients. *Ethnobot Leaf* 2009a; 13: 532–9.

Reddy BU, Seetharam YN. Antimicrobial and analgesic activities of *Trikatu Churna* and its ingredients. *Pharmacologyonline* 2009b; 3: 489–95.

Reddy PN, Lakshmana M, Udupa UV. Effect of *Praval bhasma* (Coral calx), a natural source of rich calcium on bone mineralization in rats. *Pharmacol Res* 2003; 48: 593–9.

Rekha PS, Kuttan G, Kuttan R. Effect of herbal preparation *Brahma Rasayana*, in amelioration of radiation induced damage. *Indian J Exp Biol* 2000; 38: 999–1002.

Rekha PS, Kuttan G, Kuttan R. Effect of *Brahma Rasayana* on antioxidant systems and cytokine levels in mice during cyclophosphamide administration. *J Exp Clin Cancer Res* 2001; 20: 219–23.

Renuka J, Shukla VJ, Prajapati PK. Development and validation of UV spectrophotometric method for analysis of fulvic acid – A decomposition product in *Shilajit*. *Ayurpharm – Int J Ayur All Sci* 2013; 2: 327–31.

Roshy JC, Ilanchezhian R. Experimental evaluation of *Hingusauvarchaladi Ghrita* and *Saptavartita Hingusauvarchaladi Ghrita* with special reference to their anticonvulsant activity. *Ayu* 2010; 31: 500–3.

Rout KK, Parida S, Mishra SK. Standardization of the ayurvedic formulation *Haridra Khanda* using high-performance thin-layer chromatography-densitometry. *JAOAC Int* 2008; 91: 1162–8.

Ruknuddin G, Biswajyoti P, Kumar PP, Krishnaiah AB, Basavaiah R. Anti-inflammatory and analgesic activities of *Dashanga Ghana*: An Ayurvedic compound formulation. *Int J Nutr Pharmacol Neurol Dis* 2013; 3: 303–8.

Rupapara A, Donga SB, Harisha CR, Shukla VJ. A Preliminary pharmacognostical and pharmaceutical evaluation of *Dhatri Lauha Vati* – A compound herbomineral formulation. *Int J Chem Pharm Sci* 2013; 4: 19–23.

Rupapara AV, Donga SB. Clinical evaluation of *Pandughni vati* & *Dhatri lauha vati* on *garbhini pandu* (iron deficiency anaemia) in pregnancy. *Anc Sci Life* 2012; 32: 69.

Safiullah A et al. Preparation and characterization of *Chandraprabha Vati*. *Int J Pharm Pharm Sci* 2012; 4: 55–9.

Sahoo R, Swain PK. Standardization of *Lasunadi vati*: An Ayurvedic polyherbal formulation. *Int J Pharm World* 2011; 2: 2.

Sai Prasad AJV, Ratna M, Upadhyaya BN, Appaji RR. A clinical study on the use of *Puskara Mooladi Choorna* in *Tamaka Shvasa* (Bronchial Asthma) with pulmonary function tests. *Anc Sci Life* 2010; 29: 1–5.

Saini V et al. A comparative pharmaceutical study on Ca (ion) substances of various Ayurvedic compounds. *Int J Res Ayur Pharm* 2013; 4: 586–8.

Sajith M, Sreekumar TR, Prakash V. Anthelmintic activity of *Vidangadi churna*. *Asian J Pharm Clin Res* 2013; 6: 94–5.

Saleem AM, Gopal V, Rafiullah MRM, Bharathidasan P. Chemical and pharmacological evaluation of *Karpura shilajit bhasma*, an ayurvedic diuretic formulation. *African J Trad Compl Altern Med* 2006; 3: 27–36.

Samarakoon SMS, Chandola HM, Ravishankar B, Ashok BK Gupta, Varun B. Evaluation of adaptogenic activity profile of a compound Ayurvedic formulation – *Amalakayas Rasayana*. *Indian J Trad Know* 2011a; 10: 661–7.

Samarakoon SMS, Chandola HM, Shukla VJ. Evaluation of antioxidant potential of *Amalakayas Rasayana*: A polyherbal Ayurvedic formulation. *Int J Ayurveda Res* 2011b; 2: 23–8.

Sane RT, Swati J, Gur JP, Naik SS, Sarlashkar VD, Athavale AV et al. Quality control methods for some common Ayurvedic drugs part I. *Indian drugs* 1983; 20: 351–61.

Sannd BN, Kumar A, Kumar N. To evaluate the effect of Ayurvedic drugs *Sveta parpati* with *Pasana bheda* and *Gokshura*, in the management of *mutrasmari* (urolithiasis). *J Res in Ayur Siddha* 1993; 14: 98–114.

Santhosh B, Chaudhary H, Nageswara Rao V. Nephro-protective activity of *varunadi loha* – An experimental study. *Anc Sci Life* 2012a; 32: 21.

Santhosh B, Raghuveer, Desai G, Hiremath RS, Rao N. Analytical study of *Kajjali* WSR to different samples of *Parada* (Mercury). 2012b; 1: 496. doi: 10.4172/scientific reports.496.

Santhosh B, Raghuveer, Jadar PG, Nageswara Rao V. Analytical study of *Yashada Bhasma* (Zinc Based Ayurvedic Metallic Preparation) with reference to Ancient and Modern Parameters. 2013a; 2: 582. doi: 10.4172/scientific reports.582.

Santhosh B, Raghuveer, Jadar PG, Nageswara Rao V. Screening of antioxidant activity of *Yashada Bhasma*. *Int Ayur Med J* 2013b; 1: 1–7.

Sanyal AK, Pandey BL, Goel RK. A study of the effect of *Tamra bhasma* on experimental Gastric ulcers and secretions. *Ind J Exp Biol* 1983; 21: 258–64.

Saokar RM, Sarashetti RS, Kanthi V, Savkar M, Nagthan CV. Screening of antibacterial and antifungal activity of *Gandhaka Rasayana* – an Ayurvedic formulation. *Int J Recent Trend Sci Technol* 2013; 8: 134–7.

Sarashetti RS, Simpi CC, Sandeep NM, Kanthi VG. Screening of free radical scavenging activity of *Arogyavardhini vati*. *Int J Res Ayu Pharm* 2013; 4: 555.

Saraswathy A, Suganthan J, Thomas S. High performance thin layer chromatographic finger print parameters of *Trikatu Curanam*. *Bull Medico-Ethno-Bot Res* 1998; 19: 76–85.

Sarkar PK, Prajapati PK, Choudhary AK, Shukla VJ, Ravishankar B. Haematinic evaluation of *Lauha bhasma* and *Mandura bhasma* on HgCl {2}-induced anemia in rats. *Indian J Pharm Sci* 2007; 69: 791–5.

Saroch V, Hiremath RS, Patil PA. Anticonvulsant activity of *Apasmarari rasa* – An experimental study. *Int J Ayur Med* 2012; 3.

Sathya T, Murthy B, Vardhini N. Genotoxicity evaluation of certain *Bhasmas* using micronucleus and comet assays. *Internet J Altern Med* 2008; 7: 1.

Sathya T, Murthy B, Vardhini N. Genotoxicity evaluation of certain *Bhasmas* using micronucleus and comet assays. *Internet J Altern Med* 2008; 7: 1.

Satturwar PM, Fulzele SV, Joshi SB, Dorle AK. Hepatoprotective activity of *Haridradi ghrita* on carbon tetrachloride-induced liver damage in rats. *Indian J Exp Biol* 2003; 41: 1447–51.

Satya Priya T, Deeja CR, Rama Mohan Rao G, Badari Narayana V, Suneela P, Sri Durga Ch. Clinical evaluation of *Shankhapani rasa* in the management of *kashtartava* (dysmenorrhoea). *Int J Ayur Med* 2013; 4:.

Savalgi PB, Patgiri B, Thakkar JH, Ravishankar B, Gupta VB. Evaluation of sub-chronic genotoxic potential of *Swarna Makshika Bhasma*. *Ayu* 2012; 33: 418–22.

Savrikar SS, Dole V, Ravishankar B, Shukla VJ. A comparative pharmacological investigation of three samples of '*Guduchi ghrita*' for adaptogenic activity against forced swimming induced gastric ulceration and hematological changes in albino rats. *Int J Ayurveda Res* 2010; 1: 67–72.

Savarikar SS, Barbhind MM, Halde UK, Kulkarni AP. Pharmaceutical and analytical evaluation of *triphala guggul kalpa* tablets. *J Ayurveda Integr Med* 2011; 2: 21–5.

Sawhney HL. Studies on *Malla-sindur*. *Bull Indian Med* 1974-75; 11–2.

Sawhney HL, Agrawal VK, Sharma TN. Pharmacological studies on *Malla sindur*. *J Res Indian Med* 1974; 9: 80–3.

Saxena RB. Pharmaceutical and analytical standardization of *Loknath Rasa*. *AYU* 2006; 27: 99–102.

Saxena RB, Dholakia MV, Mehta HC. Physico-chemical study of *Ashokarishta*. *J Res Ayur Siddh* 1980; 2: 279–85.

Saxena RB, Mehta HC, Dashwan MT. Kinetic measurements of disintegration of *Eladi gutika*. *Bull Medico-Ethno-Bot Res* 1985; 6: 169–76.

Saxena RB, Dhyani PL, Mehta HC. Standardisation of *Pancaguna taila*. *J Res Ayur Siddh* 1990; 11: 102–7.

Saxena RB, Mehta HC, Daswani MT, Shah KL. Standardisation of *Anu taila*. *J Res Ayur Siddh* 1992; 13: 179–82.

Schliebs R, Liebmann A, Bhattacharya SK, Kumar A, Ghosal S, Bigl V. Systemic administration of defined extracts from Withania somnifera (Indian Ginseng) and Shilajit differentially affects cholinergic but not glutamatergic and GABAergic markers in rat brain. *Neurochem Int* 1997; 30: 181–90.

Shadab M, Anupma A, Pant K. Study on antimicrobial properties of U.V. treated *Shilajit*. *Int J Toxicol Pharmacol Res* 2014; 5: 1–4.

Shah VN, Doshi DB, Shah MB, Bhatt PA. Estimation of biomarkers berberine and gallic acid in polyherbal formulation *Punarnavashtak Kwath* and its clinical study for hepatoprotective potential. *Int J Green Pharm* 2010; 4: 296–301.

Shah VN, Shah MB, Bhatt PA. Hepatoprotective activity of *Punarnavashtak kwath*, an Ayurvedic formulation, against CCl4-induced hepatotoxicity in rats and on the HepG2 cell line. *Pharm Biol* 2011; 49: 408–15.

Shailajan S, Menon SK. Polymarker based standardization of an Ayurvedic formulation, *Lavangadi Vati* using high performance thin layer chromatography. *J Pharm Res* 2011; 4: 467.

Shailajan S, Menon S, Pednekar S, Singh A. Wound healing efficacy of *Jatyadi Taila*: in vivo evaluation in rat using excision wound model. *J Ethnopharmacol* 2011a; 138: 99–104.

Shailajan S, Menon S, Yeragi M, Kelkar V. Pharmacognostic evaluation of an ayurvedic formulation *Eladi Gutika*. *Int J Green Pharm* 2011b; 5: 302–6.

Shailajan S, Sayed N, Joshi H, Tiwari B. Standardisation of an Ayurvedic formulation: *Trikatu curan* using bio analytical tools. *Int J Res Ayur Pharm* 2011c; 2: 1676–8.

Shailajan S, Chunekar S, Hande H, Matani A, Swar G, Tiwari B. Quality evaluation and standardization of an Ayurvedic formulation *Bilvadileha*: A bioanalytical approach. *J Adv Sci Res* 2013a; 4: 43–7.

Shailajan S, Menon SN, Tiwari BR, Singh AS. Standardization of *Shadbindu Taila*: An Ayurvedic oil based medicine. *Ayu* 2013b; 34: 103–7.

Shalini A et al. Antifungal activity screening and HPLC analysis of crude extract from *Tectona grandis, Shilajit, Valeriana wallachi*. *Elect J Environ Agricul Food Chem* 2009; 8: 218–29.

Shalini TV, Nagaraja TN, Balakrishna G, Satpute D Ashok, Sheshagiri S. Anti-bacterial activity of *Rasamanikya*. *Int J Res Ayur Pharm* 2011; 2: 1455–6.

Shamkuwar PB, Shahi SR. Antimotility and antisecretory effect of *Kutajarishta*: An ayurvedic antidiarrhoeal formulation. *Der Pharmacia Sinica* 2012; 3: 71–5.

Shankara MR, Murthi NS, Shastry LN. *Rasamanikya* mishrana in *Tamaka Shwasa* (Bronchial asthma). *Nagarjun* 1978; 21: 5–8.

Shanmugasundaram ER, Akbar GK, Shanmugasundaram KR. *Brahmighritham*, an Ayurvedic herbal formula for the control of epilepsy. *J Ethnopharmacol* 1991; 33: 269–76.

Sharma AK. Studies on effect of *Vatagajankusa rasa* on post polio paresis. *J Res Ayur Siddha* 1997; 18: 28–37.

Sharma AK, Borkar S. A pilot study on the effect of *Narsingha churna* in the management of *Amavata* (rheumatoid arthritis). *J Res Ayur Siddha* 2003; 24: 11–20.

Sharma AK, Sharma SM, Sharma RK. Clinical evaluation of a herbo-mineral compound drug – (*Kalpita Yoga*) in the management of *Hridroga* (Ischaemic Heart Disease). *J Res Ayur Siddh* 2002; 23: 12–21.

Sharma AK, Nepalia S, Mishra S. Evaluation of the efficacy of *Panchamrita parpati kalpa* in the management of *grahani roga* vis-a-vis ulcerative colitis. *J Res Ayur Siddha* 2006; 27: 24–40.

Sharma AK, Agarwal A, Bansal A. High performance thin layer chromatographic method for quantification of gallic acid in *arjunarishta*: an ayurvedic formulation. *Asian J Pharm Med Sci* 2011; 1: 34–39.

Sharma AK, Sharma SM, Sharma SP, Sharma AK. Management of stable angina with *Lasunadi Guggulu* – An Ayurvedic formulation. *Ann Ayur Med* 2012; 1: 15–21.

Sharma DC, Chandiramani D, Riyat M, Sharma P. Scientific evaluation of some ayurvedic preparations for correction of iron deficiency anemia. *Indian J Clin Biochem* 2007; 22: 123–8.

Sharma G, Joshi D, Aryya NC, Pandey VB. *Svarna vanga* – a short duration toxicity study. *Anc Sci Life* 1985; 5: 86–90.

Sharma K, Puri AS, Goyal HR, Sharma DN. Tamak Shwasa (Bronchial asthma) a clinical study. *J Res Indian Med* 1973; 8: 8–13.

Sharma P, Jha J, Shrinivas V, Dwivedi LK, Suresh P, Sinha M. *Shilajit*: Evaluation of its effects on blood chemistry of normal human subjects. *Anc Sci Life* 2003; 23: 114–9.

Sharma Poonam J, Jolly CI. Standardisation of the medicinal plants used in the formulation of *Abhayarishta*. *Sachitra Ayur* 1992; 44: 753–9.

Sharma PV, Singh VP. Standardisation of an ayurvedic drug: *Trivanga Bhasma*. *Anc Sci Life* 1987; 6: 148–9.

Sharma R, Amin H, Shukla VJ, Dhiman K, Galib R, Prajapat PK. Quality control evaluation of *Guduchi Satva* (solid aqueous extract of *Tinospora cordifolia* (Willd.) Miers) An herbal formulation. *Int J Green Pharm* 2013a; 7: 258–63.

Sharma R, Amin HRG, Prajapati PK. Seasonal variations in physicochemical profiles of *Guduchi Satva* (starchy substance from *Tinospora cordifolia* [Willd.] Miers). *J Ayurveda Integr Med* 2013b; 4: 93–7.

Sharma R, Kumar V, Ashok BK, Galib R, Prajapati PK, Ravishankar B. Hypoglycemic and anti-hyperglycemic activity of *Guduchi Satva* in experimental animals. *Ayu* 2013c; 34: 417–20.

Sharma RR, Bandil B. A clinical evaluation of *Rasnadi Guggulu* & *katibasti* in *katishool* (w.s.r. to lumbo-sacral disorders). *J Res Ayur Sidd* 2008; 2: 50–8.

Sharma VD, Sharma A, Kushwah HK. An indigenous approach to manage the osteoarthritis of knee joint with *lakshadi guggulu, kalka-patra bandhan* and knee traction. *Anc Sci Life* 2007; 263: 23–9.

Sharma YK, Mishra A, Mankotia R. Clinical study on the anabolic potential of *Brahma Rasayana* in geriatric patients. *Ann Ayur Med* 2012; 1: 65–70.

Shastry RA, Karadi RV, Hukkeri V. Preparation and evaluation of *Nava-Yasa churna* for hepatoprotective activity against carbon tetrachloride intoxication in albino rats. *Univ J Pharm* 2013; 2: 75–8.

Shaw BP, Jain AK, Kalita D. Clinical study of *Somraj Churna* and *Nimadi Taila* on *vicarcika* (chronic eczema). *Anc Sci Life* 1982; 1: 221–2.

Shekokar AK, Borkar K, Singh AK. To Study the efficacy of *Rasa Parpati* with *Jatyadi Ghrita Matra Basti* in the management of ulcerative colitis. *Int J Ayur Pharma Res* 2013; 1: 38–47.

Shetty P, Pushpalatha H, Bikshapathi T. Study of arogyavardhini and gandhaka rasayana in the treatment of leucoderma. *Sachitra Ayurved* 2000; 53: 438–440.

Shrivastava A, Shrivastava P, Agrawal DS, Haldar P, Sharma M. *Rasapushpa* – effect on acute and sub acute inflammation. *J Sci Innov Res* 2013; 2: 1034–9.

Shubha HS, Hiremath RS. Evaluation of antimicrobial activity of *Rasaka Bhasma*. *Ayu* 2010a; 31: 260–2.

Shubha HS, Hiremath RS. Preparation and physicochemical analysis of *Rasaka Bhasma*. *Ayu* 2010b; 31: 509–10.

Shukla RR, Verma AS, Mishra P. Efficacy of *Arogyavardhini vati* with *prakshalan* of *Kshirivriksha* in management of *karnasrava*. *Int J Ayur Med* 2013; 4: 17.

Shukla SS, Rai DK, Agrawal A, Dubey GP. Safety and efficacy of *Makardhwaj* an inorganic mercury containing Ayurvedic preparation – Study of experimentally induced renal damage in rabbits. *J Res Ayua Siddha* 1992; 1: 62–67.

Simpi RC, Sarashetti S, Patil SA. Evaluation of antimicrobial activity of *Gokshuradi Guggulu* on UTI. *J Res Ayur Sidda* 2004; 25: 67–72.

Singh JP, Tiwari SK. Clinical evaluation of *Alambushadi Ghan Vati* in *Amavata* (Rheumatoid arthritis). *J Res Ayur Siddh* 2010; 31: 85–94.

Singh JP, Antiwal M, Vaibhav A, Tripathi JS, Tiwari SK. Clinical efficacy of *Rasona Pinda* in the management of *Amavata* (rheumatoid arthritis). *Ayu* 2010; 31: 280–6.

Singh Maksoodan, Joshi Damodar, Arya NC. Studies on testicular regenerative potential of *Naga Bhasma*. *Ancient Sci Life* 1989; 9: 95–98.

Singh N, Reddy KRC, Prasad NK, Singh M. Chemical characterization of *Lauha bhasma* by X-ray diffraction and vibrating sample magnetometry. *Int J Ayur Med* 2010; 1: 3.

Singh OP, Singh R, Singh SK, Singh US. Role of *Kankayana vati, Triphala churna* and *Kasishadi taila* in the management of *arsha* (anorectal piles). *J Res Ayur Siddh* 2005; 26: 59–65.

Singh OP, Padhi MM, Das B, Deep VC, Hazra J, Alam MM, Tewari NS, Rao MM. Clinical evaluation of *Arogyavardhini vati, Kaishore guggulu* and *Chakramardakera taila* in the management of *kitibh* (psoriasis). *J Res Ayur Siddh* 2007; 28: 61–71.

Singh OP, Padhi MM, Das B, Deep VC, Hazra J, Alam MM, Tewari NS, Rao MM. Clinical evaluation of *Kanchnar guggulu* and *Gokshuradi guggulu* in the management of manifested cases of *Slipada* (Filariasis). *J Res Ayur Siddh* 2008; 29: 39–47.

Singh R, Sharma G, Aryya NC, Joshi D, Kumar N. Toxicity study and testicular regeneration property of *swarna vanga*. *Anc Sci Life* 2012; 32: 11.

Singh RK, Ahmad M, Wafai ZA, Seth V, Moghe VV, Upadhyaya P. Anti-inflammatory effects of *Dashmula*, an Ayurvedic preparation, versus Diclofenac in animal models. *J Chem Pharm Res* 2011; 3: 882–8.

Singh RK, Banerjee R, Upadhyay S, Mitra A, Hazra J. Toxicological evaluation of *Panchakola Avaleha*, an Ayurvedic classical formulation, in albino rats. *Ayu* 2012; 33: 303–6.

Singh SK, Rai SB. Detection of carbonaceous material in *naga bhasma*. *Indian J Pharm Sci* 2012; 74: 178–83.

Singh SK, Rajoria K. Clinical study on *Lakshadi guggulu* and *Panchatikta Ksheer Vasti* in osteoarthritis of knee joint. *J Ayush* 2014; 3: 68–79.

Singh SK, Chaudhary AK, Rai DK, Rai SB. Preparation and characterization of a mercury based Indian traditional drug *Ras-Sindoor*. *Indian J Trad Know* 2009; 8: 346–57.

Singh SK, Gautam DN, Kumar M, Rai SB. Synthesis, characterization and histopathological study of a lead-based Indian traditional drug: *naga bhasma*. *Indian J Pharm Sci* 2010; 72: 24–30.

Sivakumar V, Sivakumar S. Effect of an indigenous herbal compound preparation '*Trikatu*' on the lipid profiles of atherogenic diet and standard diet fed *Rattus norvegicus*. *Phytother Res* 2004; 18: 976–81.

Soni A, Patel K, Gupta SN. Clinical evaluation of *Vardhamana Pippali Rasayana* in the management of *Amavata* (Rheumatoid Arthritis). *Ayu* 2011; 32: 177–180.

Sridhar BN. The role of *Trikatu yoga* in the management of *pratisyaya*. *Aryavaidyan* 2001; 14: 154–8.

Srilakshmi D, Anand T, Khanum F, Kumar TP, Sreelakshmi C. *In-vivo* toxicity evaluation of *Shila Sindoor*. *Int J Ayur Pharma Research* 2013; 1: 24–30.

Srivastava A. A clinical experiment to explore the systemic toxicology of *Mrga Srnga Bhasma*. *Sach Ayur* 1992; 44: 600.

Sriwastava NK, Shreedhara CS, Aswatha Ram HN. *In-vitro* free radical scavenging activity of *Ajmodadi Churna* – A polyherbal formulation. *J Pharm Res* 2010a; 3: 1467–70.

Sriwastava NK, Shreedhara CS, Aswatha Ram HN. Standardization of *Ajmodadi churna*, a polyherbal formulation. *Pharmacog Res* 2010b; 2: 98–101.

Sruthi CV, Sindhu A. A comparison of the antioxidant property of five Ayurvedic formulations commonly used in the management of *vata vyadhis*. *J Ayurveda Integr Med* 2012; 3: 29–32.

Sud S, Sudheendra H, Sujatha K, Sekhar Reddy P. A comparative Pharmaceutico-analytical evaluation of *Rasamanikya* prepared with three different methods. *Int J Res Ayur Pharm* 2011; 2: 1651–4.

Sud S, Reddy PS, Sujatha K, Honwad S. A comparative toxicological study of *Rasamanikya* prepared with three different methods. *Ayu* 2013; 34: 309–15.

Sudhakar P, Gopalakrishna HN, Swati B, Shreyasi C, Raghavendra B, Preethi G Pai, Pai MRSM. Effect of *Tantupasana* on electrical and chemical induced seizures in mice. *J Pharm Res* 2010; 3: 1178–80.

Sumantran VN, Kulkarni AA, Harsulkar A, Wele A, Koppikar SJ, Chandwaskar R, Gaire V, Dalvi M, Wagh UV. Hyaluronidase and collagenase inhibitory activities of the herbal formulation *Triphala guggulu*. *J Biosci* 2007; 32: 755–61.

Surapanenia DK, Adapaa SRSS, Preetia K, Ravi Tejaa G, Veeraragavanb M, Krishnamurthy S. Shilajit attenuates behavioral symptoms of chronic fatigue syndrome by modulating the hypothalamic–pituitary–adrenal axis and mitochondrial bioenergetics in rats. *J Ethnopharmacol* 2012; 143: 91–9.

Suresh P, Joshi D, Gode KD, Chakravarty BK. Effect of *Swarna vanga* on *madhumeha* in albino rats. *Anc Sci Life* 1988; 8: 30–7.

Swain PK, Sahoo R, Patra S, Kulshrestha MK. Evaluation of an Ayurvedic formulation: *Apamarga Kshara*. *World J Pharm Pharmaceut Sci* 2013; 2: 2726–9.

Swamy GK, Bhattathiri PPN. *Vatari Guggulu* and *Maharasnadi Quatha* in the management of *Amavata* – A clinical study. *J Res Ayur and Siddh* 1998; 19: 41–8.

Swamy GK, Bhattathiri PPN, Narayana A. Clinical evaluation of *Haridra Khanda* in the management of *Sitta Pitta*. *J Res Ayur Sidd* 1999; 20: 20–8.

Tambekar DH, Dahikar SB. Screening antibacterial activity of some bhasma (metal-based medicines) against enteric pathogen. *Recent Res Sci Technol* 2010; 2: 59–62.

Tamboli S. A clinical evaluation of *vasant kusumakara rasa* in *prameha upadrava* with special reference to diabetic neuropathy and retinopathy. University of Mumbai, 2001.

Tank ZG, Joorawon PR, Choudhary AK, Pande D. A clinical study of *Shankha bhasma* alone and along with *Amalaki churna* in the management of *Amlapitta*. *Sach Ayur* 2005; 57: 527–33.

Tanna Ila, Chandola HM, Joshi JR. Clinical efficacy of *Mehamudgara vati* in type 2 diabetes mellitus. *Ayu* 2011; 32: 30–9.

Tanna Ila R, Aghera HB, Ashok BK, Chandola HM. Protective role of *Ashwagandharishta* and flax seed oil against maximal electroshock induced seizures in albino rats. *AYU* 2012; 33: 114–8.

Teli P, Jadhav J, Kanase A. Comparison of *Abhrak Bhasma* and silicon dioxide efficacy against single dose of carbon tetrachloride induced hepatotoxicity in rat by evaluation of lipid peroxidation. *American J Pharm Health Res* 2014; 14: 123–135.

Tewari VP, Tewari KC, Joshi P. An interpretation of Ayurvedic findings on *Shilajit*. *J Res Ind Med* 1973; 8: 53–58.

Thabrew M Ira, Dharmasiri MG, Senaratne L. Anti-inflammatory and analgesic activity in the polyherbal formulation: *Maharasnadi Quatha*. *J Ethnopharmacol* 2003; 85: 261–7.
Thakur SN, Srinivas C, Deshpande PJ. Spectroscopic analysis of *Yashada Bhasma* (zinc salt). *Anc Sci Life* 1986; 5: 240–2.
Thangapazham RL, Sharma A, Gaddipati JP, Singh AK, Maheshwari RK. Inhibition of tumor angiogenesis by *Brahma Rasayana* (BR). *J Exp Ther Oncol* 2006; 6: 13–21.
Thanki K, Bhatt N, Shukla VD. Effect of *kshara basti* and *nirgundi ghana vati* on *amavata* (rheumatoid arthritis). *Ayu* 2012; 33: 50–3.
Thirunavukkaras SV, Upadhyay L, Venkataraman S. Effect of *Manasamitra vatakam*, an Ayurvedic formulation, on aluminium-induced neurotoxicity in rats. *Trop J Pharm Res* 2012; 11. http://www.ajol.info/index.php/tjpr/article/view/74672
Thundiparambil CJ, Poly S, Kulkarni PV, Joseph R, Ilanchezhian R. Preliminary analytical study of *Gandarvahasthadi Kwath* – An Ayurvedic polyherbal formulation. *Ayurpharm Int J Ayur Alli Sci* 2012; 1: 41–5.
Tiwari HS, Patgiri B, Prajapati PK. A comparative study on *Pravala Mula Bhasma* and *Pravala Shakha Bhasma* in *Amlapitta*. *J Res Ayur Siddh* 2008; 29: 123–8.
Tiwari P, Patel RK. Evaluation of diuretic potential of *Draksarista* prepared by traditional and modern methods in experimental rats. *Pharmacologyonline* 2011; 3: 566–72.
Tiwari P, Ramarao P, Ghosal S. Effects of *Shilajit* on the development of tolerance to morphine in mice. *Phytother Res* 2001; 15: 177–9.
Tiwari R, Pandya DH, Baghel MS. Clinical evaluation of *Bilvadileha* in the management of irritable bowel syndrome. *Ayu* 2013; 34: 368–72.
Tiwari SK, Pattanshetty JK, Pushpalatha H. Standardisation of *Bhringaraja Taila*. *J Res Ayur Siddh* 1993; 14: 83–7.
Tripathi R, Rathore AS, Mehra BL, Raghubir R. Physico-chemical study of *Vaikranta bhasma*. *Anc Sci Life* 2013; 32: 199–204.
Tripathi YB, Singh VP. Role of *Tamra bhasma* and Ayurvedic preparation in the management of lipid peroxidation in liver of albino rats. *Ind J Expl Biol* 1996; 34: 66–70.
Tripathi YB, Shukla S, Chaturvedi S. Antilipidperoxidative property of *Shilajit*. *Phytother Res* 1996; 10: 269–73.
Tripathy R, Panda M. Clinical study on the management of fractures with *Abha-Guggulu*. *J Res Ayur Sidd* 2009; 30: 1–8.
Trivedi VP, Nesamony S, Sharma VK. A clinical study of the antitussive and antiasthmatic effects of *Vibhitakyadi churna* in the cases of *kasa-swasa*. *J Res Ayur Siddha* 1982; 3: 1–8.
Ullah M Obayed, Hamid K, Rahman K Ashfaqur, Choudhuri MSK. Effect of *Rohitakarista*, an Ayurvedic formulation, on the lipid profile of rat plasma after chronic administration. *Biol Med* 2010; 2: 26–31.
Ullah MO, Shrestha T, Munira TS, Choudhuri MS. Pharmacological and toxicological studies of an ayurvedic formulation ("*Lauhasava*") on the biological system of rats and mice. *Pak J Biol Sci* 2008; 11: 2013–7.
Vahalia MK, Thakur KS, Nadkarni K, Sangle VD. Chronic toxicity study for *Tamra Bhasma* (A Generic Ayurvedic Mineral Formulation) in laboratory animals. *Recent Res Sci Technol* 2011; 3: 76–79.
Vani S, Tiwari SK. A clinical study on the efficacy of *Satavari Mandur* in *Parinamasoola* vis a vis peptic ulcer. *Aryavaidyan* 2002; 15: 189–191.
Varma S, Shamsia S, Thiyagarajan OS, Vidyashankar S, Patki PS. *Yashada bhasma* (Zinc calx) and *Tankana* (Borax) inhibit Propionibacterium acne and suppresses acne induced inflammation *in vitro*. *Int J Cosmetic Sci* 2014; 36: 361–8.
Vasanthakumar KG, Pattanshetty JK, Bikshapathi T, Vijayalakshmi B. Standardization studies on *arjunarishta*. *Bull Medico-Ethno-Botanical Res* 2003; 24: 97–102.
Vasanthakumari V, Shyamala Devi C. Evaluation of antidiabetic effect of an ayurvedic drug, '*tarakeswara rasa*' in rats. *Indian J Exp Biol* 1997; 35: 909–11.
Vasudevan TN, Khorana ML. Standardization of Ayurvedic preparation – Assay of *Kutajarishta* and the crude drugs. *Indian J Pharm* 1965; 27: 96.
Velmurugan C, Vivek B, Wilson E, Bharathi T, Sundaram T. Evaluation of safety profile of black *Shilajit* after 91 days repeated administration in rats. *Asian Pac J Trop Biomed* 2012; 2: 210–4.
Velpandian T, Mathur P, Sengupta S, Gupta SK. Preventive effect of *Chyavanprash* against steroid induced cataract in the developing chick embryo. *Phytother Res* 1998; 12: 320–3.
Vemula SK, Chawada MB, Thakur KS, Vahalia MK. Antiulcer activity of *Amlapitta Mishran* suspension in rats: A pilot study. *Anc Sci Life* 2012; 32: 112–5.

Verma N. Chorea cured with Ayurvedic preparation. *J Res Ayur Siddh* 1985; 6: 101–4.

Verma PR, Prasad CM. Standardization and bioavailability of ayurvedic drug *lauha bhasma* part-1 physical and chemical evaluation. *Anc Sci Life* 1995a; 15: 129–36.

Verma PR, Prasad CM. Standardization and bioavailability of ayurvedic drug *lauha bhasma* – part-II comparative bioavailability studies. *Anc Sci Life* 1995b; 15: 140–4.

Vishwapal SH, Handa KL. Some inorganic preparations of Indian indigenous medicine-*shank*, *parval*, *shiringi* and *godanti bhasmas*. *Indian J Pharmacy* 1958; 20: 182–4.

Vishwesh BN, Dwivedi M. A clinical study to evaluate the role of *Shivagutika* in cystic ovarian disease. *Ayurpharm-Int J Ayur All Sci* 2014; 3: 1–5.

Vite M, Nangude S, Chugh N. Standardization of Ayurvedic formulation: *Jatyadi ghrita*. *Int J Pharm Pharm Sci* 2013; 5: 378–85.

Vora DK, Banerjee SK, Chhabra GS. Method development for estimation of alcohol in Ayurvedic formulations using gas chromatography. *Bull Pharm Res* 2012; 2: 34–7.

Wadekarv M, Gogte V, Khandagale P, Prabhunev A. Comparative study of some commercial samples of *Naga bhasma*. *J Anci Sci Life* 2004; 23: 1–9.

Wadher SJ, Puranik M, Yeole PG, Lokhande CS. Determination of ethanol in *abhayarishta* by gas chromatography. *Indian J Pharm Sci* 2007; 69: 152–4.

Wavare R, Yadav R, Sheth S, Sawant R. A pharmaceutical approach of *Manikya Bhasma* towards its standardisation. *J Pharm Sci Innov* 2014a; 3: 82–6.

Wavare R, Yadav R, Sheth S, Sawant R. A pharmaceutical approach of *Manikya Pisthi* towards its standardisation. *Global J Res Med Plant Indigen Med* 2014b; 3: 105–11.

Yadav JS, Thakur S, Chadha P. *ChyawanprashAwaleha*: a genoprotective agent for bidi smokers. *Int J Human Genet* 2003; 3: 33–8.

Yadav KD, Reddy KRC. Standardization of *Brahmi Ghrita* with special reference to its pharmaceutical study. *Int J Ayur Med* 2012; 3: 16–21.

Yadav KD, Reddy KR, Kumar V. Study of *Brahmi Ghrita* and piracetam in amnesia. *Anc Sci Life* 2012; 32: 11–5.

Yadav KD, Reddy KR, Agarwal A. Preliminary physico-chemical profile of *Brahmi Ghrita*. *Ayu* 2013; 34: 294–6.

Yadav KD, Reddy K, Kumar V. Acute and sub-chronic toxicity study of *Brahmi ghrita* in rodents. *Int J Green Pharm* 2014; 8: 18–22.

Yadav SS, Galib, Ravishankar B, Prajapati PK, Ashok BK, Varun B. Anti-inflammatory activity of *Shirishavaleha*: An Ayurvedic compound formulation. *Int J Ayurveda Res* 2010; 1: 205–7.

Yadav SS, Galib, Prajapati PK, Ashok BK, Ravishankar B. Evaluation of immunomodulatory activity of "*Shirishavaleha*" – An Ayurvedic compound formulation in albino rats. *J Ayurveda Integr Med* 2011; 2: 192–6.

Zala U, Kumar V, Chaudhari AK, Ravishankar B, Prajapati PK. Anti-inflammatory and analgesic activities of *Panchtikta Ghritha*. *Ayurpharm-Int J Ayur All Sci* 2012; 1: 187–92.

NOTIFICATIONS

A

Notification regarding Shelf life or expiry date for Ayurvedic medicines

MINISTRY OF HEALTH AND FAMILY WELFARE

[Department of Ayurveda, Yoga and Naturopathy, Unani, Siddha & Homoeopathy] (Ayush)

24th November, 2005

G.S.R. 691(E).—The following draft of certain rules further to amend the Drugs and Cosmetics Rules, 1945 which the Central Government proposes to make, in exercise of the powers conferred by clause (a) of sub-section (2) of section 33-N of the Drugs and Cosmetics Act 1940 (23 of 1940), is hereby published as required by the said Section for the information of all persons likely to be affected thereby and a notice is hereby given that the said draft rules will be taken into consideration after the expiry of a period of thirty days from the date on which copies of the Official Gazette in which this notification is published, are made available to the public;

(1) These rules may be called the Drugs and Cosmetics (Amendment) Rules, 2005.
(2) They shall come into force on the date of their publication in the Official Gazette.
In the Drug and Cosmetics Rules, 1945;
after rule 161A, the following rule shall be added, namely:-

"161B:—The date of expiry of Ayurveda, Siddha and Unani (ASU) medicines shall be conspicuously displayed on the label of container or package of Ayurvedic, Siddhas and Unani drugs, after which they shall not be in circulation. Shelf-life for Ayurveda, Siddha & Unani (ASU) medicines shall be as follows:

Shelf life or expiry date for Ayurvedic medicines

S.No.	Name of the Group of Ayurvedic Medicine	Shelf Life &/Expiry Date
1.	Churna/Kwatha Churna	1 year
2.	Gutika (Varti-Gutti/Pills/Tablets except Gutika with Rasa)	2 years
3.	(i) Gutika Tablet containing Kastha aushadhi	2 years

Contd....

Contd.

	(ii) Gutika/Tablet containing Kasth oushadi and Rasa, Uprasa, Metalic Bhasmas, and Guggulu.	5 years
4.	Rasausadhies	No expiry date *
5.	Asava Arista	No expiry date *
6.	Avaleha	2 years
7.	Guggulu	5 years
8.	Mandura-Lauha	10 years
9.	Ghrita	1 year and 6 months
10.	Taila	2 years
11.	Arka	1 year
12.	Dravaka/Lavana/Ksara	5 years
13.	Lepa Churna	1 year
14.	Lepa Guti	3 years
15.	Lepa Malahar (Ointment)/Liniment/Gels	2 years
16.	Varti	2 years (one time use)
17.	Ghana Vati	2 years
18.	Kupipakva Rasayan	No expiry date *
19.	Parpati	No expiry date *
20.	Sveta parpati	2 years
21.	Pisti & Bhasma	No expiry date *
22.	Svarna, Rajata, Lauha, Mandura, Abhraka bhasma, Godanti, Shankha Bhasma, etc.	No expiry date *
23.	Naga Bhasma, Vanga Bhasma, Tamra Bhasma **	5 years **
24.	Capsules made of soft gelatin (depending upon the content material) for Kashtha aushadhi	2 years
25.	Capsules of hard gelatin (depending upon the content material) – containing Kashtha aushadhi with Rasa, Bhasma, Parada-Gandhak	3 years
26.	Syrup/liquid oral	3 years
27.	(Karna/Nasa Bindu) Ear/Nasal drops	2 years
	Eye drops	1 year
28.	Khand/Granule/Pak	2 years
29.	Dhoopans-Inhalers	2 years
30.	Pravahi Kwatha	One year 6 months

* Item at Sr. No. 4, 5, 19, 20, 22, 23 have very long shelf life. They became more efficacious with the passage of time. For keeping a record, 10 year period should be mandatory for such items.

** These Bhasmas start solidifying after five years. Therefore, they need one or two 'Puta' again before being used in the dosage form.

B

Notification regarding for mandatory testing of heavy metals by Ministry of AYUSH

F.No.K-11020/5/97-DCC (AYUSH)
GOVERNMENT OF INDIA MINISTRY OF HEALTH & FAMILY WELFARE

(Department of Ayurveda, Yoga & Naturopathy)

Central Government vide its Order of even number dated 14.10.2005 has made testing for heavy metals, namely, Arsenic, Lead, Mercury and Cadmium mandatory for export purposes in respect of every batch of purely herbal Ayurveda, Siddha and Unani medicines by every licensee. Representations have been received from Ayurveda Drugs Manufacturers Association (ADMA) that the exporters may find it difficult to print the certificate Heavy metals within permissible limit on the container of the medicines as container and label are to be approved by the regulatory authorities of the importing countries and it may not be possible to get revised containers and labels approved by the regulatory authorities of the importing countries by 1st January, 2006.

In view of the above difficulties brought to the notice of the Central Government, the above mentioned order of even number dated 14.10.2005 is hereby partially modified as follows:

In case any manufacturer of purely herbal Ayurveda, Siddha and Unani medicines finds it difficult to display on the container of purely herbal Ayurveda, Siddha and Unani medicines to be exported the words Heavy metals within permissible limit from 1st January, 2006 in view of the regulatory requirements of the importing country, the manufacturer shall submit batchwise testing reports in respect of every purely herbal Ayurveda, Siddha and Unani medicine to be exported from approved laboratories certifying that the medicine contains heavy metals within permissible limits as laid down by the Central Government vide the order of even number dated 14.10.2005. The test report of an approved laboratory will be an essential part of the consignment papers to be submitted by any exporter from 1st January, 2006 which will be examined by the representative of the DCG(I) at the Airport/port of shipment. The manufacturers/exporters of purely herbal Ayurveda, Siddha and Unani medicines shall be responsible for the genuineness of the test report of the approved laboratory submitted along with other consignment papers at the Airport/Port of Shipment. All GLP/NABL accredited laboratories are approved by the Central Government for the above mandatory testing. The list of approved laboratories and the permissible limits of heavy metals have been posted on the website of the Department of AYUSH, Ministry of Health and Family Welfare, Government of India www.indianmedicine.nic.in.

The order issued by the Central Government under section 33 EEB of the Drugs and Cosmetics Act, 1940 dated 14.10.2005 as modified hereby shall come into force from 1st January, 2006. All exporters of purely herbal Ayurveda, Siddha and Unani medicines are hereby directed to either conspicuously display on the container of purely herbal Ayurveda, Siddha and Unani medicines to be exported the words Heavy metals within permissible limit or furnish the above mentioned certificate from an approved laboratory along with other consignment papers. It will be the responsibility of the representative of DCG(I) deployed at the Airport/port of shipment to examine and ensure that all exporters of purely herbal Ayurveda, Siddha and Unani medicines comply with this Order w.e.f. 1st January, 2006.

C
Notification regarding for mandatory testing of heavy metals by Ministry of Finance
F.No. 450/163/2005-Cus.IV
Government of India
Ministry of Finance Department of Revenue
Central Board of Excise and Customs

March 21st, 2006

The undersigned is directed to bring to your kind attention that the order issued by the Central Government under section 33 EEB of the Drugs and Cosmetics Act, 1940 dated 14.10.2005 as modified, has come into force from 1st January, 2006. All exporters of purely herbal Ayurveda, Siddha and Unani Medicines (AYUSH) are required to either conspicuously display on the container of purely herbal Ayurveda, Siddha and Unani medicines to be exported the words "Heavy metals within permissible limits" or furnish the prescribed certificate from an approved laboratory along with other consignment papers. It is the responsibility of the representative of DCG(I) deployed of the Airport/port of shipment to examine and ensure that all exporters of purely herbal Ayurveda, Siddha and Unani medicines comply with the Order w.e.f. 1st January, 2006. The copy of above order dated 14.10.2005 as amended by order dated 14.12.2005 is enclosed. Details are available at website www.indianmedicine.nic.in.

In view of above, you are requested to issue suitable instructions to field formations under your jurisdiction.

D

Notification regarding Order Ingredient List
F.No. K-11020/5/97 – DCC (AYUSH)
GOVERNMENT OF INDIA
MINISTRY OF HEALTH & FAMILY WELFARE
(DEPARTMENT OF AYURVEDA, YOGA & NATUROPATHY, UNANI, SIDDHA AND HOMOEOPATHY)

March 14, 2006

Whereas the Govt. Of India in the Ministry of Health and Family Welfare, Department of Ayurveda, Yoga & Naturopathy, Unani, Siddha and Homoeopathy (AYUSH) had issued an order of even number on October 10, 2005 in exercise of the powers conferred under section 33p of the Drugs and Cosmetics Act, 1940, Govt. of India in the Ministry of Health and Family Welfare, Department of Ayurveda, Yoga & Naturopathy, Unani, Siddha and Homoeopathy (AYUSH) directing the State Licensing Authorities of Ayurveda, Siddha and Unani (ASU) drugs to ensure full compliance by all ASU drug manufactures of the Provision of Rule 161 (1) and (2) relating to displaying on the label of the Container or Package of an Ayurveda, Siddha and Unani drugs, the true list of all the ingredients (official and botanical names) used in the manufacture of the preparation together with the quantity of each of the ingredients incorporated therein.

Whereas representatives of Ayurveda, Siddha and Unani drugs manufactures have made a representation to the Central Government for modification in the above order to provide for printing of only the official names of all the ingredients used in the manufacture of the preparation together with the quantity of each of the ingredients incorporated therein and dispense with the requirement of printing corresponding botanical names.

Whereas representatives of Ayurveda, Siddha and Unani drugs manufacture companies have also made a representation to Central Government that some grace period be provided to them to comply with the requirement of indicating the true list of all the ingredients together with the quantity of each of the ingredients incorporated therein so that the medicines already in the market are not adversely affected.

In view of the above representations, order issued by the Govt. of India in the Ministry of Health and Family Welfare, Department of Ayurveda, Yoga & Naturopathy, Unani, Siddha and Homoeopathy (AYUSH) of even number dated October 10, 2005 is hereby modified to provide that State Drug Licensing Authorities of ASU drugs shall ensure full compliance by all ASU drug manufactures of Provisions of Rule 161 (1) and (2) relating to displaying on the label of the container or in a leaflet to be inserted in the package of ASU drugs. Both classical as well as patent Proprietary, official names of all the ingredients, used in the manufacture of that drug together with the quantity of each of the ingredients incorporated therein. However, the manufactures shall continue to provide corresponding botanical names to the state Drug Licensing Authorities as part of the license application. Further that

State Drug Licensing Authorities shall enforce the order in respect of all ASU drugs with affect from 1.7.2006. The remaining parts of the order of even number dated October 10, 2005 remain unchanged.

E
Notification regarding Maintaining of records of raw material
AYUSH Notification No.: G.S.R. 512(E) (09-Jul-08) Drug and Cosmetics (First Amendment) Rules, 2008

1. Whereas the draft of certain rules further to amend the Drugs and Cosmetics Rules, 1945 was published as required by section 33N of the Drugs and Cosmetics Act, 1940 (23 of 1940), in the Gazette of India, Extraordinary, dated the 19th October, 2006, vide Number GSR 651(E) inviting objections and suggestions from persons likely to be affected thereby and notice was given that the said draft will be taken into consideration after the expiry of a period of forty-five days from the date on which copies of the Official Gazette containing the said notification were made available to the public;

   And whereas, the said Gazette was made available to the public on the 18th October, 2006;

   And whereas, objections and suggestions received from the public on the said draft rules Drugs and Cosmetics Act, 1940 (23 of 1940) have been considered by the Central Government;

   Now, therefore, in exercise of the powers conferred by section 33-N of the Drugs and Cosmetics Act, 1940 (23 of 1940) the Central Government, after consultation with the Ayurveda, Siddha and Unani Drugs Technical Advisory Board, hereby makes the following rules further to amend the Drugs and Cosmetics Rules, 1945, namely:

   (1) These rules may be called the Drug and Cosmetics (First Amendment) Rules, 2008.
   (2) They shall come into force from the date of their publication in the Official Gazette.

2. In the Drug and Cosmetics Rules, 1945 (herein referred to as the said rules), after rule 157, the following rule shall be inserted, namely:-

   "157A. Maintaining of records of raw material used by licensed manufacturing unit of Ayurveda, Siddha and Unani drugs in the preceding financial year. Each licensed manufacturing unit of Ayurveda or Siddha or Unani drugs shall keep a record of raw material used by each licensed manufacturing unit of Ayurveda, Siddha or Unani drugs the case may be in the performa given in Schedule TA in respect of all raw materials utilized by that unit in the manufacture of Ayurveda or Siddha or Unani drugs in the preceding financial year, and shall submit the same by the 30th day of June of the succeeding financial year to the State Drug

Licensing Authority of Ayurveda, Siddha and Unani drugs and to the National Medicinal Plants Board or any agency nominated by the National Medicinal Plant Board for this purpose."

3. In the said rules, after Schedule T, the following Schedule shall be inserted.

F

Notification regarding Drugs and Cosmetics (6th Amendment) Rules, 2010

MINISTRY OF HEALTH AND FAMILY WELFARE

(Department of Ayurveda, Yoga and Naturopathy, Unani, Siddha and Homoeopathy) (AYUSH)

10th August, 2010

G.S.R. 663(E).—Whereas the draft of certain rules further to amend the Drugs and Cosmetics Rules, 1945 was published, vide notification of the Government of India in the Ministry of Health and Family Welfare, number G.S.R. 377(E), dated 3rd May, 2010, in the Gazette of India, Extraordinary, inviting objections and suggestions from persons likely to be affected thereby before the expiry a period of Forty Five days from the date on which copies of the Official Gazette containing the said notification were made available to the public;

And whereas, the said Gazette was made available to the public on the 4th May, 2010;

And whereas, objections and suggestions received from the public on the said draft rules have been considered by the Central Government;

Now, therefore, in exercise of the powers conferred by section 33-N of the Drugs and Cosmetics Act, 1940 (23 of 1940) the Central Government, hereby makes the following rules further to amend the Drugs and Cosmetics Rules, 1945, namely:-

1. These rules may be called the Drugs and Cosmetics (6th Amendment) Rules, 2010. They shall come into force on the date of their publication in the Official Gazette.
2. In the Drugs and Cosmetics Rules, 1945 (herein after referred to as the said rules), after rule 158-A, the following rules shall be inserted, namely:-

158(B) Guidelines for issue of license with respect to Ayurveda, Siddha or Unani drugs.

I. (A) Ayurveda, Siddha Unani Medicines under section 3(a):

Ayurveda, Siddha or Unani drugs includes all medicines intended for internal or external use for or in the diagnosis, treatment, mitigation or prevention of disease or disorder in human beings or animals, and manufactured exclusively in accordance with the formulae described in the authoritative books of Ayurvedic, Siddha and Unani Tibb system of medicine, as specified in the First Schedule;

(B) Patent or Proprietary medicine under section 3(h);

(i) In relation to Ayurvedic, Siddha and Unani Tibb system of medicine of all formulations containing only such ingredients mentioned in the formulae described in the authoritative books of Ayurveda, Siddha or Unani Tibb system of medicines specified in the First Schedule, but does not include a medicine which is administered by parenteral route and also a formulation included in the authoritative books as specified in clause (a).

(ii) **Balya/Poshak/Muqawi/Unavuporutkal/positive health Promoter** formulations having ingredients mentioned in books of First Schedule of the Drugs and Cosmetics Act and recommended for promotional and preventive health.

(iii) **Saundarya Prasadak (Husane afza)/Azhagh-sadhan** formulation having ingredients mentioned in Books of First Schedule of the Drugs and Cosmetics Act and recommended for oral, skin, hair and body care.

(iv) **Aushadh Ghana (Medicinal plant extracts – dry/wet)** extract obtained from plant mentioned in books of First Schedule of the Act including Aqueous or hydro-alcohol.

II. (A) For issue of license to the medicine with respect to Ayurvedic, Siddha and Unani, the conditions relating to safety study and the experience or evidence of effectiveness shall be such as specified in columns (5) and (6) of the Table given below:

S. No.	Category	Ingredient(s)	Indications(s)	Safety Study	Experience/Evidence of Effectiveness	
1	2	3	4	5	6	
					Published Literature	Not Required
1.	(A) Ayurveda, Siddha and Unani drugs given in 158-B as referred in 3(a)	As per text	As per text	Not Required	Required	Not Required
2.	(B) Any change in dosage form of Ayurveda Siddha and Unani drugs as described in section 3(a) of the Drugs and Cosmetics Act, 1940	As per text	As per text	Not Required	Required	Not Required
3.	(C) Ayurveda, Siddha and Unani drugs referred in 3(a) to be used for new indication	As per text	New	Not Require	IF Required	Required

(B) For issue of license with respect to Patent or Proprietary medicine. The condition relating to Safety studies and experience or evidence of effectiveness shall be specified as follows:

S. No.	Category	Ingredient(s)	Indications(s)	Safety Study	Experience/Evidence of Effectiveness	
					Published Literature	Not Required
1	2	3	4	5	6	
1.	Patent or Proprietary medicine	As per text	Textual rationale	Not Required	Of Ingredients	*Pilot study as per relevant protocol for Ayurveda, Siddha and Unani drugs
2.	Ayurveda Siddha, Unani drug with any of the ingredients of Schedule E(1) of The Drugs and Cosmetics Act, 1940	As per text	Existing	Required	Required	Required

(III) For issue of license with respect to Balya and Poshak medicines, the person who applied for license is required to submit the following:

(i) Photo-copy of the textual reference of ingredients used in the formulation as mentioned in the book of 1st schedule;
(ii) Conduct safety studies in case the product contains of any of the ingredients as specified in the Schedule E(1), as per the guidelines for evaluation of Ayurveda Siddha and Unani Drugs formulations;
(iii) For textual indications the safety and effectiveness study is not required.

(IV) For issue of license with respect to Saundarya Prasadak (Husane afza/Azhagu Sodhan) the person who applied for license is required to:

(i) Submit photo-copy of the textual reference of ingredients used in the formulation as mentioned in the book of 1st schedule;
(ii) Conduct safety studies, in case the formulation contains of any of the ingredients as specified in the Schedule E(1), as per the guidelines for evaluation of Ayurveda, Siddha and Unani formulation;
(iii) For textual indications the safety and effectiveness study is not required.

(V) For issue of license with respect to medicine Aushadh Ghana [extract of medicinal plant] (dry or wet).

S. No.	Category	Ingredient(s)	Indications(s)	Safety Study	Experience/Evidence of Effectiveness	
1	2	3	4	5	6	
					Published Literature	Not Required
1.	(A) Aqueous	As per Text	As per text	Not Required	Not Required	Not Required
2.	(A1) Aqueous	As per Text	New indication	Not Required	Not Required	Required
3.	(B) HydroAlcohol	As per Text	As per Text	Not Required	If Required	Not Required
4.	(B1) HydroAlcohol	As specified	New Indication**	Required	If Required	Required
5.	Other than Hydro/ HydroAlcohol	As specified	As specified	Required Acute, Chronic, Mutagenicity and Teratogenicity	If Required	Required

* The standard protocol will also include concept of Anupan, Prakriti & Tridosh, etc. published by Central Research Councils Ayurveda, Siddha, Unani and other Government/Research Bodies.
** New indication means that which is other than what is mentioned in 1st schedule of Drugs & Cosmetics Act, 1940.
Foot Note: The Principal rules were published in Official Gazette vide notification No. F.28-10/45-H(I) dated the 21st December, 1945 and the last amended vide No. GSR 602(E), dated 19-7-2010.

G

Notification regarding Exemption from Excise Duty on Medicaments used in Ayurvedic system under Notification No. 75/94-CE

F.No. 345/19/95-TRU

Government of India
Department of Revenue
(Tax Research Unit), New Delhi

1. Ayurvedic system of medicines in accordance with the formulae given in the authoritative books specified in the first Schedule to the Drugs and Cosmetic Act and sold under a generic name are fully exempt from excise duty under notification No. 75/94-CE, dated the 29th March, 1944. The scope of the exemption on Ayurvedic, Unani, Siddha, Homeopathic and Bio-chemics systems was clarified by the Board through a Circular, F.No. B30/1/94-TRU, dated the 29th

March, 1994. It was made clear that the benefit of exemption is applicable so long as the medicine is sold by name described in such books or pharmacopoeia. It is immaterial whether the manufacturer puts his mark, logo, monogram, symbol, etc. in marketing such medicines.

2. It has now been brought to the notice of the Board that ayurvedic medicine manufacturers are facing certain difficulties in availing of exemption from excise duty in cases where some other ingredients are also added so as to increase the shelf life or to make it in a convenient form or to be palatable. Such ingredients are reported to be not having any therapeutic value. However, the benefit of exemption from excise duty is being denied on the ground that such medicaments are not strictly being manufactured as per the standard text-books which is a condition prescribed under the said notification.

3. The matter has been examined. There is no dispute that the medicaments remain ayurvedic medicines and are sold as Ayurvedic medicines under generic name as stated in the standard textbooks even if preservatives or binding agents, etc. are added but the question is whether in view of the existing condition specified under the notification the concessional rate of duty will be available or not. A view has been expressed that as long as the ingredients are added in the ayurvedic medicaments which do not have any therapeutic value and the formulation is otherwise prepared as per the prescribed textbooks, it may not be proper to deny the benefit of full exemption to such of those ayurvedic medicaments from excise duty under notification No. 75/94-CE. In this context, the Ayurvedic Siddha and Unani Drugs Technical Advisory Board under the Ministry of Health and Family Welfare were consulted, who has opined that significant changes have taken place in the process of making of ayurvedic medicines and preservatives, inner excipients like starch, lactose and willing agents like sodium laurel sulphate are nowadays widely used. They are in favour of allowing use of preservatives by the manufacturers of ayurvedic medicines.

4. Having regard to the above, the Board is of the view the benefit of exemption under S.No. 3 of notification No. 75/94-CE should not be denied to ayurvedic medicines if they are manufactured in accordance with the formulae prescribed in the said authoritative textbooks but contain preservatives, the assessing officers may obtain a certificate to be produced by the manufacturers from the appropriate Drug Authorities that the drug satisfies the following conditions:-
   (a) The medicament has been prepared in accordance with the formulae prescribed in the authoritative textbook; and
   (b) The ingredients added other than those prescribed in the authoritative textbooks, should not have any therapeutic value.
   Needless to say that for availing the benefit of exemption, the ayurvedic medicament should be sold under the generic name as specified in the authoritative textbooks.

5. Pending cases on this issue, if any, may be disposed off accordingly.

FORMS

FORM 24-D

Application for the grant/renewal of a licence to manufacture for sale of Ayurvedic drugs

1. I/We ………………………….. of …………………………………..hereby apply for the grant/renewal of a licence to manufacture Ayurvedic drugs on the premises situated at…………………………………………
2. Names of drugs to be manufactured (with details).
3. Names, qualification and experience of technical staff employed for manufacture and testing of Ayurvedic drugs ………………………………………..
4. A fee of rupees……………………..has been credited to the Government under the head of account……………………………and the relevant Treasury Challan is enclosed herewith.

Date…………   Signature……………..

(Applicant) NOTE—The application should be accompanied by a Plan of the premises.

FORM 24-E

Application for grant or renewal of a loan licence to manufacture or sell Ayurvedic Drugs

1. I/We*……………………………………………….of**………………………. hereby apply for the grant/renewal of a loan licence to manufacture Ayurvedic Drugs……………………………………..on the premises situated at…………… ………………………………………………………………………………….. C/o ***…………………………………………………..
2. Names of drugs to be manufactured (with details).
3. The names, qualifications and experience of technical staff actually connected with the manufacture and testing of Ayurvedic (including Siddha) or Unani drugs in the manufacturing premises.
4. I/We enclose,
    a) A true copy of a letter from me/us to the manufacture concern whose manufacturing capacity is intended to be utilized by me/us.

b) A true copy of a letter from the manufacturing concern that they agree to lend the services of their competent technical staff, equipment and premises for the manufacture of each item required by me/us and that they shall maintain the registers of raw materials and finished products separately in this behalf.

c) Specimen of labels, cartons of the drugs proposed to be manufactured.

A fee of Rs..has been credited to Government under the head of account...........................and the relevant Treasury Challan is enclosed herewith.

Date............   Signature................

* Enter here the name of the proprietor, partners or Managing Director as the case may be.

** Enter here the name of the applicant firm and the address of the principal place of business.

*** Enter here the name and address of the manufacturing concern where the manufacture will be actually carried out and also the licence number under which the letter operates.

FORM 25-D

Licence to manufacture for sale of Ayurvedic drugs

No. of Licence..............................

1. ....is/are hereby licensed to manufacture the following Ayurvedic drugs on the premises situated at................................under the direction and Supervision of the following technical staff: —
   a) Technical staff (Name).
   b) Names of drugs (each item to be separately specified).
2. The licence shall be in force from...............................to...............
3. The licence is subject to the conditions stated below and to such other conditions as may be specified in the Rules for the time being in force under the Drugs and Cosmetics Act, 1940.

Date............   Signature................
                                                                                                  Designation..............

FORM 25-E

Loan Licence to manufacture for sale Ayurvedic Drugs

1. Number of Licence..
2. ....of...........................is hereby granted a loan licence to manufacture for sale Ayurvedic (including Siddha) or Unani drugs, on the premises situated at.. C/o.....................................under the direction and supervision of the following expert technical staff.
   a) Technical staff
   b) Names of drugs (each item to be separately specified)

3. The licence shall be in force from.......................to...........................
4. The licence is subject to the conditions stated below and to such other conditions as may be specified in the Rules for the time being in force under the Drugs and Cosmetics Act, 1940.

Date of issue............   Signature..................
                                                                         Designation...............

FORM 26-D

Certificate of renewal of licence to manufacture for sale of Ayurvedic drugs

1. Certified that licence No......................granted on the......................to Shri/Messers..............................for the manufacture of Ayurvedic/Siddha/Unani drugs at the premises situated at...........has been renewed from..............................to.............................
2. Name of technical staff..
3. Names of drugs (each item to be separately specified).

Date............   Signature..................
                                                             Designation...............

FORM 26-E

Certificate of renewal of loan licence to manufacture for sale of Ayurvedic/Siddha or Unani Drugs

1. Certified that Loan Licence No...................................granted on the....................to...........................for the manufacture of Ayurvedic/Siddha or Unani drugs at the premises situated at........... ....C/o...........................has been renewed from..to.......................................
2. Name of technical staff...

Date............   Signature...............
                                                           Designation...............

FORM 26-E-1

Certificate of Good Manufacturing Practices (GMP) to manufacture of Ayurvedic drugs

Certified that manufacturing unit licensee, namely ....situated at .... State .... Licence No. .... comply with the requirements of Good Manufacturing Practices of Ayurveda-Siddha-Unani drugs as laid down in Schedule T of the Drugs and Cosmetic Rules, 1945.
This certificate is valid for a period of three years.

Date ....   Signature..................
Place ....   Designation...............

Licensing Authority for Ayurveda/Siddha/Unani Drugs.

TABLES

Table 1. Chemical analysis of *Godanti Bhasma*.

	%
Iron	Negligible
Ca as CaSO$_4$	29.88% w/w
Sulphate	Minimum (69.07% w/w) Maximum (67.081% w/w)
Mangenese	Maximum (0.623% w/w)
Acid-insoluble ash	Minimum (27.68% w/w) Maximum (99.90% w/w)
Loss on drying	Minimum (0.127% w/w) Maximum (7.127% w/w)

Source: CCRIMH.

Table 2. Chemical analysis of *Shankh Bhasma*.

	%
Calcium (as calcium)	Minimum (34.400% w/w) Maximum (42.278% w/w)
Ca as CaCO$_3$	Maximum (54.50% w/w)
Sulphate	Maximum (1/295% w/w)
Mangenese	Maximum (0.23% w/w)
Acid-insoluble ash	Maximum (0.719% w/w)
Loss on drying	Maximum (0.06% w/w)

Source: CCRIMH.

Table 3. Chemical analysis of *Mrg Shrng Bhasma*.

	%
Iron as FeO$_3$	Maximum (0.032% w/w)
Calcium (as calcium)	Minimum (37.25% w/w) Maximum (38.06% w/w)
Phosphate	Maximum (17.49% w/w)
Magnesium	Maximum (3.4% w/w)
Ash amount	Minimum (98.76% w/w) Maximum (99.80% w/w)
Acid-insoluble ash	Minimum (0.950% w/w) Maximum (0.194% w/w)
Loss on drying	Maximum (0.438% w/w)

Source: CCRIMH.

Table 4. Comparative chemical analytical data of selected calcium preparations.

Parameter	Praval bhasma	Shahnka bhasma	Shukti bhasma	Varatika bhasma	Godanti bhasma	Kukkutandatvak bhasma
Ash value % w/v (at 450°C)	93.198	95.922	97.185	95.381	97.126	97.376
Ph of 10% w/v solution in water	12.52	12.38	12.45	12.49	12.43	12.44
Acid soluble matter % w/v	99.862	99.582	99.288	99.732	93.873	97.713
Calcium as CaO % w/w	72.067	63.491	82.891	67.484	45.990	76.559

Source: Vinod et al. *Int J Res Ayur Pharm* 2013; 4: 586–8.

Table 5. Chemical analysis of *Hartala Bhasma*.

Appearance	White Yellowish
Odour	Odourless
UV Analysis	0.025% w/v solution in water at 200 nm to 380 nm of UV range (a) 229 nm = 0.907 (b) 283 nm = 0.052
PH of 5% w/v suspension	11.24
Ash value w/w %	99.10
Acid insoluble ash w/w %	2.06
Loss of drying	0.555
Arsenic as As w/w %	10.28
Sulphur as S w/w	20.914

Source: Kumar et al. *Int J Ayur Pharm Res* 2013; 1: 36–46.

Table 6. Chemical analysis of *Hartala godanti Bhasma*.

Appearance	White
Odour	Odourless
UV Analysis	No absorption at 200 nm to 380 nm of UV range
PH of 5% w/v suspension	10.46
Ash value w/w %	99.802
Acid insoluble ash w/w %	5.275
Loss of drying (at 110 degree C) w/w %	2.00
Arsenic as As w/w %	1.625
Sulphur as S w/w	8.574
Calcium as Ca w/w %	31.970

Source: Kumar et al. *Int J Ayur Pharm Res* 2013; 1(3): 36–46.

Table 7. Chemical analysis of *Makardwaz*.

Free sulphur	Nil
Sulphur	Minimum (11.5% w/w) Maximum (14.0% w/w)
Mercury	Minimum (72.2% w/w) Maximum (86.0% w/w)
Gold	absent

Source: CCRIMH.

Table 8. Chemical analysis of *Rasa Sindura*.

Free sulphur	Minimum (1.4% w/w) Maximum (7.7% w/w)
Sulphur	Minimum (13.2% w/w) Maximum (13.8% w/w)
Mercury	Minimum (82.7% w/w) Maximum (87.1% w/w)
Ash value	Nil

Source: CCRIMH.

Table 8. Chemical analysis of *Rasa Karpura*.

Free sulphur	Nil
Sulphur	Nil
Mercury	Minimum (64.3% w/w) Maximum (79.6% w/w)
Chloride as Cl	Maximum (24.55% w/w)
Sodium as Na	Trace

Table 9. Chemical analysis of *Sameer Panag Rasa*.

Free sulphur	Minimum (1.4% w/w) Maximum (3.3% w/w)
Sulphur	Minimum (20.03% w/w)
Mercury	Minimum (12.0% w/w) Maximum (17.2% w/w)
Arsenic	Minimum (30.28% w/w) Maximum (32.50% w/w)

Source: CCRIMH.

Table 10. Chemical analysis of *Rasa Pushpa*.

Free sulphur	Nil
Sulphur	Nil
Mercury HgO	Maximum (17.73% w/w)
Iron as Fe_2O_3	Minimum (0.721% w/w) Maximum (11.870% w/w)
Chloride as Cl	Maximum (21.16% w/w)
Acid insoluble ash	Maximum (0.63% w/w)
Loss on drying	Maximum (17.1% w/w)

Source: CCRIMH.

Table 11. Chemical analysis of *Suvarna Vanga*.

Free sulphur	Nil
Sulphur	Minimum (12.7% w/w) Maximum (35.2% w/w)
Mercury	Maximum (1.0% w/w)
Tin as SnS_2	Minimum (33.7% w/w) Maximum (64.4% w/w)
Tin as SnO_2	Minimum (63.45% w/w) Maximum (80.41% w/w)

Source: CCRIMH.

Table 12. X-ray diffraction of the *Swarna makshika Bhasma*.

Compounds	d' value	R. Intensity
Fe_2O_3	2.51 2.95	100 32.4
FeS_2	2.69 1.69	82.2 24.3
CuS	3.03 3.16 1.45	50.6 26.6 12.0
SiO_2	3.67 1.60	31.9 8.9

Note: N.B.: d' value-'d' space; R. Intensity-Relative intensity.
Source: Mohaptra and Jha *Int J Ayurveda Res* 2010; 1: 82–6.

Table 13. Elemental analysis of *Swarna Bhasma* by EDAX.

Element	SB_1 (%)	SB_2 (%)
As	9.95	9.56
Nb	0.56	0.72
Au	88.10	91.2

Table 14. Chemical analysis of *Trivanga Bhasma*.

Name of the Test	Result
Description	
Color	Yellowish
Odour	Nil
Identification	Positive results for Pb, Zn, Sn
Loss on drying	0.1801%
Total ash content	99.57%
Acid insoluble ash content	94.3654%
Water soluble ash content	1.2509%
Qualitative analysis	
Iron	Positive
Calcium	Positive
Free metal content	Negative
Assay for elements	
Sn	38.26%
Zn	17.067%
Pb	31.67%
Si	3.54%
Mn	356.98%
Cl	0.34%
Ca	3.98%
Fe_2SO_3	3.15%
Al	1.68%
Ni	161.084 ppm*
Cu	679.69 ppm*
Cd	106.67 ppm*
Ayurvedic specifications	
Lusterless	Non-luster
Rekhapurantva	Positive

*ppm = parts per million
Source: CCRAS.

Table 15. Chemical composition of *Hiraka Bhasma* by EDX.

Constituents	DHP*	UNZ*	AR*	UMP*	KG*
Carbon	7.17	7.55	9.27	6.92	32.02
Oxygen	27.45	26.15	25.49	28.87	34.42
Sodium	1.04	0.84	1.00	0.68	2.46
Magnesium	4.35	4.55	3.19	1.19	2.15
Aluminum	3.48	2.40	4.48	3.47	6.76
Silicon	5.64	6.02	6.66	6.80	10.61
Phosphorus	1.06	-	1.02	0.74	-
Sulfur	2.51	2.56	1.51		-
Potassium	2.71	3.10	4.74	2.21	0.26
Calcium	4.67	5.95	5.77	3.18	2.81
Chromium	-	-	-	1.79	-
Iron	39.92	41.00	32.87	42.47	7.56

Percentage (weight) of the constituents
Source: Kadam, Jawale and Wadekar Indian Science Congress 2007.

Table 16. Chemical analysis of *Karpura shilajit bhasma*.

Element	Amount (% w/w)
Calcium	24.6044
Aluminium	0.0966
Magnesium	0.3666
Iron	0.7511
Nitrogen	3.124
Aloin	1.25

Source: Saleem et al. *African J Trad Complem Altern Med* 2006; 3: 27–36.

Table 17. *Asava* with high content of alcohol as base.

S.No.	Name of the Preparation	Maximum No. Size of Packing
1	*Kapurasava*	15 ml
2	*Ahiphenasava*	15 ml
3	*Margamadasava*	15 ml

Source: http://www.plimism.nic.in/Protocol_For_Testing.pdf.

Table 18. Preparations containing self-generated alcohol.

S.No.	Name of the Preparation	Maximum Content of Alcohol (w/w)	Maximum Size of Packing
1	*Mrtsanjivanisura*	16.00%	30 ml
2	*Mahadraksasava*	16.00%	120 ml

Source: http://www.plimism.nic.in/Protocol_For_Testing.pdf.

Table 19. Standards to be complied with in manufacture for sale or distribution of ASU Drugs.

S.No.	Class of Drugs	Standards to be Complied With
1	Drugs included in the standards for identity, Purity and Ayurvedic Pharmacopoeia	The standards for identity, purity and strength as given in the editions of Ayurvedic Pharmacopoeia of India for the time being in force.
2	*Asava* and *Arista*	The standards for identity, purity and strength as given in the editions of Ayurvedic Pharmacopoeia of India for the time being in force.

Source: http://www.plimism.nic.in/Protocol_For_Testing.pdf.

Table 20. Chemical analysis of *Rasa Parpati*.

Free sulphur	Minimum (38.2% w/w) Maximum (54.0% w/w)
Sulphur	Minimum (45.80% w/w) Maximum (48.73% w/w)
Mercury	Minimum (35.60% w/w) Maximum (47.10% w/w)
Iron	Trace

Source: CCRIMH.

Table 21. Chemical analysis of *Lauha Parpati*.

Free sulphur	Minimum (43.8% w/w) Maximum (47.0% w/w)
Sulphur	Minimum (48.0% w/w) Maximum (50.0% w/w)
Mercury	Minimum (17.5% w/w) Maximum (25.8% w/w)
Iron as Fe_2O_3	Minimum (17.07% w/w) Maximum (17.26% w/w)

Source: CCRIMH.

Table 22. Chemical analysis of *Swarna Parpati*.

Free sulphur	Minimum (35.5% w/w) Maximum (38.0% w/w)
Sulphur	Minimum (37.6% w/w) Maximum (44.7% w/w)
Mercury	Minimum (26.0% w/w) Maximum (38.4% w/w)
Gold as Au	Minimum (6.0% w/w) Maximum (10.3% w/w)

Source: CCRIMH.

Table 23. Chemical analysis of *Bola Parpati.*

Free sulphur	Minimum (21.7% w/w) Maximum (22.5% w/w)
Sulphur	Minimum (24.4% w/w) Maximum (24.7% w/w)
Mercury	Minimum (25.20% w/w) Maximum (27.30% w/w)
Calcium	Trace
Iron	Trace
Resin content	14.7% w/w

Source: CCRIMH.

Table 24. Chemical analysis of *Panchamita Parpati.*

Free sulphur	Minimum (44.7% w/w) Maximum (47.5% w/w)
Sulphur	Minimum (47.4% w/w) Maximum (53.6% w/w)
Mercury	Minimum (18.6% w/w) Maximum (37.8% w/w)
Copper as Cu	Minimum (1.1% w/w) Maximum (2.1% w/w)
Aluminium as Al_2O_3	Maximum (31.5% w/w)
Iron as Fe_2O_3	Minimum (1.4% w/w) Maximum (18.0% w/w)
Calcium as Ca	0.814% w/w
Magnesium as Mg	Trace
Chloride as Cl	Trace
Sulphate	1.18% w/w
Phosphate as PO_4	0.38% w/w
Acid insoluble	3.3% w/w

Source: CCRIMH.

Table 25. X-ray diffraction of *Praval bhasma.*

d' Value	Relative Intensity
15.60	100.00
12.15	47.90
6.34	31.77
4.89	32.00
3.62	38.60
2.03	68.35
1.69	59.00
1.45	59.75

Source: Mishra et al. *J Integr Med* 2014; 12: 52–8.

Table 26. Chemical analysis of *Abhraka Bhasma*.

Iron as Fe_2O_3	Minimum (20.07% w/w) Maximum (27.60% w/w)
Calcium	Minimum (1.65% w/w) Maximum (4.86% w/w)
Sodium	Maximum (1.76% w/w)
Potassium	Maximum (5.36% w/w)
Phosphate as PO_4	Maximum (1.71% w/w)
Aluminium as Al_2O_3	Minimum (3.80% w/w) Maximum (18.55% w/w)
Magnesium	Minimum (4.80% w/w) Maximum (7.55% w/w)
Silica	Maximum (36.74% w/w)
Acid insoluble	Maximum (60.72% w/w)
Ash value	Minimum (18.71% w/w) Maximum (100.00% w/w)
Acid insoluble ash	Minimum (47.7% w/w) Maximum (63.7% w/w)
Loss on drying	Maximum (0.43% w/w)

Source: CCRIMH.

Table 27. Chemical analysis of *Swarna Makshika Bhasma*.

Free sulphur	Minimum (1.43% w/w) Maximum (6.31% w/w)
Sulphur	Maximum (3.33% w/w)
Iron as Fe_2O_3	Minimum (20.07% w/w) Maximum (27.60% w/w)
Calcium	Maximum (1.625% w/w)
Sodium	Maximum (0.370% w/w)
Potassium	Maximum (5.36% w/w)
Sulphate	Maximum (3.00% w/w)
Copper	Maximum (17.2% w/w)
Ferric oxide	Maximum (25.0% w/w)
Ferrous oxide	Maximum (5.7% w/w)
Phosphate as PO_4	Maximum (1.101% w/w)
Silica	Maximum (3.8% w/w)
Acid insoluble	Maximum (11.73% w/w)
Ash value	Minimum (12.10% w/w) Maximum (18.20% w/w)
Acid insoluble ash	Minimum (21.20% w/w) Maximum (31.18% w/w)

Source: CCRIMH.

Table 28. Physical constants of *Somnathi Tamra Bhasma*.

Ash value	51.40%
Acid insoluble ash	7.16%
Water insoluble ash	19.96%
Loss on drying	1.4
pH	5.33
Specific gravity	1.002

Source: Honwad et al. *J Biol Sci Opn* 2014; 2: 396–401.

Table 29. Physico-chemical parameters of *Tamra Bhasma* samples.

Parameters (% w/w)	TB1	TB2	TB3	Avg
Loss on drying	1.10	0/96	0.86	0.97
Ash value	97.79	96.58	97.89	97.42
Acid insoluble ash	2.35	2.64	1.89	2.29
CS_2 soluble extractive	1.32	1.38	1.25	1.33
Water soluble extractive	0.76	0.78	0.84	0.80

Source: Jagtap et al. *Anc Sci Life* 2012; 31: 164–70.

Table 30. Elemental analysis of raw ruby, *Manikya Bhasma* (MB1 & MB2).

Elements	Raw Ruby	MB1	MB2
Al	51.2	43.8	43.2
Cu	21.6	115.2	81.6
Fe	1600	34996	33316
Cr	2884	3884	3748
K	274	2644	2684
Na	52	4048	4460
Ca	174.4	15820	15316
Mn	ND	506	503
Mg	1004	8828	8660
As	ND	ND	ND
S	ND	ND	ND

Source: Ramesh et al. *J Pharm Sci* 2014; 3: 82–6.

Table 31. Elemental composition of pearl and *Mukta Shukti Bhasma* by EDAX analysis.

Elements	Pearl	*Mukta Shukti Bhasma*
Ca	43.26 ± 0.07	40.22 ± 0.05
C	16.09 ± 1.12	16.42 ± 1.4
O	40.65 ± 0.08	42.80 ± 0.09

Source: Dubey et al. *Songklanakarin J Sci Technol* 2009; 31: 501–10.

Table 32. Elemental composition of *Varatika Bhasma*.

Sl. No.	Name of the Parameter	Raw *Varatika*	*Varatika bhasma*
1.	Ash (%)	4.13	2.06
2.	Organic Carbon (%)	(%) 0.52	1.09
3.	Total Nitrogen (%)	0.56	0.72
4.	Total Phosphorus (%)	0.36	0.62
5.	Total Potassium (%)	3.26	3.49
6.	Total Sodium (%)	1.06	1.36
7.	Total Calcium (%)	15.63	19.32
8.	Total Magnesium (%)	8.56	8.43
9.	Total Sulphur (%)	0.78	0.94
10.	Total Zinc (ppm)	1.56	1.48
11.	Total Copper (ppm)	0.52	0.42
12.	Total Iron (ppm)	102.0	113.6
13.	Total Manganese (ppm)	17.55	19.62
14.	Total Boron (ppm)	0.08	0.06
15.	Total Molybdenum (ppm)	0.02	0.03

Source: Devanathan et al. *Int J Curr Pharm Res* 2010; 2: 12–6.

Table 33. Analytical specifications of *Kantakaryavaleha*.

Specific gravity	1.32
Loss on drying	28.5%
Ash	4.6%
Acid insoluble ash	1.1%
Fat	15.4%
Reducing sugar	12.3%
Non-reducing sugar	20.7%
Total alkaloids	0.19%

Source: Alam et al. *B M E B R* 1983; 3: 97–102.

Table 34. Analytical specifications of *Narikela Khanda*.

Specific gravity	1.113
Loss on drying	8.7%
Ash	1.802%
Acid insoluble ash	0.212%
Fat	45.2%
Reducing sugar	5.62%
Non-reducing sugar	34.9%
Total alkaloids	0.134%
Calcium	0.036%
Phosphates	0.39%

Source: Alam et al. *B M E B R* 1983; 3: 97–102.

Table 34. India's export of medicinal plants to major markets during the period 2002–03 to 2004–05.

Country	2002–03	2003–04	2004–05	Percent Growth
USA	135.63	111.92	98.51	(–)11.98
Japan	30.06	18.57	11.86	(–)36.13
Germany	13.37	13.26	11.55	(–)12.90
France	6.76	8.13	6.76	(–)16.85
UK	17.00	11.45	11.12	(–)2.88
China	6.73	7.04	8.66	23.01
Hong Kong	8.24	10.51	7.51	(–)33.30
Pakistan	6.44	7.29	9.20	26.20
UAE	9.88	5.72	10.47	83.04
Taiwan	8.46	8.43	5.77	(–)31.55
Total (All India)	334.17	302.11	263.08	(–)12.92

Source: Compiled from the data of DGCI&S, Monthly statistics of India foreign trade: Exports & re-exports March 2003 & 2004 issues, Kolkata.

Table 35. Medicinal plants being exported from India.

Botanical Name	Part of the Plant
Acorus calamus	Rhizome
Adhatoda vasica	Whole plant
Cassia angustifolia	Leaf & pod
Colchicum luteum	Rhizome & Seed
Picrorhiza kurroa	Root
Rauvolfia serpentina	Root
Rheum emodi	Rhizome
Swertia chirayita	Whole plant
Whole plant	Root
Zingiber officinale	Rhizome

Source: CDRI 1998.

Table 36. Medicinal plants being imported from India.

Botanical Name	Native Name
Cuscuta epithymum	Aftimum vilaiyti
Glycyrrhiza glabra	Mulethi
Lavandula stoechas	Ustukhuddus
Smilax china	Chobehini
Smilax ornata	Ushba

Source: CDRI 1998.

Table 37. Medicinal species included by state amendments in section 2(4) (a) of 1927 Indian Forestry Act.

State	Species
Gujarat	*Rauwolfia serpentina*, Karaya gum
Karnataka	Sandalwood oil, rosha grass and oil, *Phyllanthus embica, Terminalia chebula, Terminalia belerica, Capparis mooni*
Kerala	Gum, fibres and roots of sandalwood and rosewood
Maharashtra	Rosha grass including oil, *Rauwolfia serpentina*
Orissa	Roots of Patargaruda, sandalwood, tamarind
Tamil Nadu	Sandalwood
Tamil Nadu	Sandalwood
Uttar Pradesh	Gum, Chiraunji

Source: Jha 1990.

LISTS

List 1. LIST OF APPROVED DRUG TESTING LABS (Under Rule -160 A to J of the Drugs and Cosmetics Rule 1945)

1. M/S Varun Herbals Pvt. Ltd., 5-8-293/A, Mahesh Nagar, Chirag Ali Lane, Hyderabad.
2. M/S Sipra Labs Pvt. Ltd., 407, Nilgiri, Adiya Enclave, Ameetpet, Hyderabad-500 038.
3. Captain Srinivasa Murti Drug Research Institute for Ayurveda (CCRAS), Arumbakkam, Chennai-600 106.
4. M/s Sowparnika Herbal Extracts & Pharmaceuticals Pvt. Ltd., No. 31-A/2A, North Phase, SIDCO Industrial Estate, Chennai-600 098.
5. Regional Research Laboratory (CSIR), Canal Road, Jammu Tavi, Jammu-180 001.
6. ARBRO Pharmaceuticals Ltd., 4/9, Kirti Nagar Industrial Area, New Delhi-110 015.
7. Shriram Institute for Industrial Research, 14 & 15 Sathyamangala Industrial Area, Whitefield Road, Bangalore-560 048.
8. Bangalore Test House, 65/20th Main Morenhalli, Vijayanagar, Bangalore.
9. FRLHT, 74/2 Jarakabande Kaval, Post Attur Via Yelahanka, Bangalore-560 064.
10. M.S. Ramaiah Drugs and Allied Products Testing Laboratories, M.S. Ramaiah Nagar, M.S.R.I.T. (POST), Bangalore-560 054.
11. Ozone Pharmaceuticals Ltd. (Analytical Lab), MIE, Bahadurgarh, Dist. Jajjhar, Haryana-124 507.
12. M/S Standard Analytical & Research Laboratories, 358/4, Laxmibai Nagar, Industrial Estate, Kilamaidan, Indore, M.P.-452 004.
13. M/S Anusandhan Analytical & Biochemical Research Laboratory Pvt. Ltd., Indore, 435, M.G. Road, Shiv Vilas Place, Rajwada, Indore, M.P.-452 004.
14. M/S Choksi Laboratories Ltd., 6/3, Manoramganj, Indore, M.P.-452 001.
15. M/S Quality Control Laboratory, Plot No. 17, Malviya Nagar, Bhopal, M.P.-462 003.
16. J.R.D. Tata Foundation for Research in Ayurveda and Yoga Sciences, Arogyadham (Deendayal Research Institute) Chitrakoot, Satna, M.P.-485 331.
17. J.R.D. Tata Foundation for Research in Ayurveda and Yoga Sciences, Arogyadham (Deendayal Research Institute) Chitrakoot, Satna, M.P.-485 331.
18. M/s Sitharam Ayurved Pharmacy Ltd., Nedumpuzha, Thrissur, Kerala.
19. Vaidya Rathnam, Aushadhshala, Ollur, Thrissur.

20. Arya Vaidyashala Kotakkal, Malapuram.
21. M/s Natural Remedies Pvt. Ltd., Bangalore.
22. Drug Testing Laboratory, Niper, Mohali, Chandigarh.
23. M/s Standard Analytical Laboratory Pvt. Ltd. 69, Functional Industrial estate, Patparganj, Delhi-92.
24. Shilpacham Manufacturers of Ayurvedic Pharmaceutical Products, 47-D estate Fort, Laxmibai Nagar, Indore-452006.
25. Laboratories Service Division, Sargam Metals Pvt. Ltd. Chennai.
26. M/s Amol Pharmaceutical Pvt. Ltd., Sanganer Jaipur, Rajasthan.
27. M/s Charak Testing Laboratory, 32 & 67, Evergreen Industrial Estate, Shakti Mills Lane, Mahalaxmi, Mumbai-400 011.
28. M/s Delhi Test House, A-62/3, G.T. Karnal Road, Institutional Area, opposite Hans Cinema, Azadpur, Delhi-33.

Source: www.indianmedicine.nic.in/index3.asp?sslid=848&subsublinkid=231&lang=1.

List 2. LIST OF ASSISTED STATE GOVT. DRUG TESTING LABORATORIES FOR AYURVEDA SIDDHA UNANI DRUGS

1. Chief Supdt., Govt. Drug Testing Laboratory, Govt. Indian Medicine Pharmacy (Ayu.), Kattedan, Hyderabad, 500077.
2. Dy. Director, Govt. Drug Testing Laboratory, Govt. Central Pharmacy, Jayanagar, I Block near Ashoka pillar, Bangalore, Karnataka.
3. Dean, Govt. Drug Testing Laboratory, Govt. Ayurvedic and Unani Pharmacy, Vazirabad, Nanded, Maharashtra-431602.
4. Senior Scientific Officer & Govt. Analyst Food & Drug Laboratory, Near polytechnic, Vadodara, Gujrat.
5. Manager, Ayurved Drug Testing Laboratory, Govt. Ayurvedic Pharmacy, Pushkar Road, Ajmer, Rajasthan.
6. Govt. Analyst Laboratory, For Ayurvedic & Unani Medicine, Takait Rai Talab, Mohan Road, Lucknow, UP.
7. Incharge, Govt. Drug Testing Laboratory, Research Institute in Indian System of Medicine, Joginder nagar, Distt. Mandi, H.P.-176120.
8. Principal, Govt. Drug Testing Laboratory, Rishikul State Ayurvedic College, Haridwar, Uttaranchal.
9. Senior Research Officer (Chemistry), Incharge Drug Testing Laboratory, Ayurvedic Research Institute, Drug Standardization Unit, Poojappura, Thiruvananthapuram.
10. Dr. Rabinarayan Acharya, Scientific Officer & Head, State Drug Testing & Research Laboratory (ISM) Govt. Ayurvedic Hospital Campus, Nagarwartangi, P.O. BJB Nagar. Bhubaneshwar, Orissa-751014.
11. Superintendent, Govt. Drug Testing Laboratory (ISM), Arignar Anna Govt. Hospital of Indian Medicine Complex, Arumbakkam, Chennai, (T.N.).
12. Director, State Pharmacopoeial Laboratory & Pharmacy for Indian Medicine, Kalyani, Nadia. Ph. 50626281.

13. Govt. Analyst and Incharge, Drug Testing Laboratory, Govt. Ayurvedic Pharmacy Campus, Amkho Lashkar, Gwalior M.P.-474009.
14. Superintendent, Drug Testing Laboratory, Govt. Ayurvedic Pharmacy, GE Road, Raipur, Chhatisgarh.
15. Controller Combined Food & Drug Laboratory, Patoli, Mangotrian, Jammu-180007.
16. Incharge, State Drug Testing Laboratory for ISM Drugs, Govt. Ayurvedic College, Guwahati.
17. Deputy Drug Controller, State Drug Testing Laboratory (ISM), Aushadh Niyantran Bhawan, Pt. Nehru Office Complex, Agartala-799006, Tripura.
18. Programme Officer (ISM&H), Incharge Drug Testing Laboratory, Central Medical Store, Zamabawk, Aizawl, Mizoram.
19. Govt. Analyst and Incharge, Govt. Drug Testing Laboratory (ISM), Food & Drug Laboratory, Pesteur Institute, Shillong, Meghalaya-793001.
20. Deputy Director (ISM&H), Incharge, Government Drug Testing Laboratory (ISM), Neheralagrum, Itanagar, Arunachal Pradesh.
21. Asstt. Director ISM&H, Directorate of Health Services, Govt. Drug Testing Laboratory for AYUSH, Kohima, Nagaland.
22. Superintendent, Govt. Drug Testing Laboratory (Ayurveda) Govt. Central Pharmacy & Store, Old Press Road, Patiala, Punjab.
23. Principal, Incharge, Drug Testing Laboratory (ISM), Sri Krishna Govt. Ayurveda College & Hospital, Kurukshetra, Haryana.
24. Deputy Director (AYUSH), Govt. Drug Testing Laboratory, Ranchi, Jharkhand.
25. Incharge, Govt. Drug Testing Laboratory, Ayurveda and Unani Pharmacy campus, Patna, Bihar.
26. Directorate of Health Services, State Drug Testing Laboratory for ASU&H drugs, Chander, Sikkim.
27. Incharge, Drug Testing Laboratory for ASU drugs, SASTRA, Thanjavur, Tamil Nadu.

Source: http://indianmedicine.nic.in/showfile.asp?lid=328.

List 3. DELHI DECLARATION ON TRADITIONAL MEDICINE FOR THE SOUTH-EAST ASIAN COUNTRIES

The Delhi Declaration is as follows:

A.

We, the Health Ministers of South-East Asian countries, representing the Governments of Bangladesh, Bhutan, India, Nepal, Minister of Indigenous Medicine, Sri Lanka, and Vice Minister of Health, Timor-Leste, and the representatives of DPR Korea, Indonesia, Myanmar, Maldives and Thailand, met in New Delhi during the "International Conference on Traditional Medicine for South-East Asian Countries", and we –

1) Recalled the importance given at the International Conference on Primary Health Care at Alma Ata in 1978 for inclusion of access to Traditional Medicine in the planning and implementation of health care;
2) Noted the progress of Traditional Medicine in the countries of South East Asia Region, specifically after the World Health Organization (WHO) brought out the strategy for Traditional Medicine 2002–2005;
3) Considered the importance of various resolutions of the World Health Assembly (WHA) and of the South East Asia Regional Committee for promoting Traditional Medicine and Medicinal Plants, specifically WHA56.31, WHA62.13 and SEA/RC56/R6;
4) Appreciated the diversity and richness of Traditional Medical Systems, their courses of study, status of research & development, regulatory frameworks and medicinal flora in the South-East Asian countries;
5) Recognized that Traditional Medicine and Traditional Medicine Practitioners have substantial potential to contribute for improving health outcomes in various countries of the world;
6) Acknowledged the fact that traditional medicine is culturally acceptable, generally available, affordable and widely used in various countries for the treatment of diseases;
7) Noted the fact that for millions of people often living in rural areas in different countries, traditional medicine is a significant source of health care;
8) Recognized the potential of traditional medicine in providing primary health care; and
9) Expressed the need for sharing of experience and knowledge for securing reliance on Traditional Medicine for public health benefits.

B. DECLARATION

In the light of the above, we hereby agree for cooperation, collaboration and mutual support amongst the South-East Asian Countries in all spheres of Traditional Medicine in accordance with national priorities, legislations and circumstances, and specifically agree to make collaborative efforts aiming at the following:

I. To promote National policies, strategies and interventions for equitable development and appropriate use of traditional medicine in the health care delivery system;
II. To develop institutionalized mechanism for exchange of information, expertise and knowledge with active cooperation with WHO on traditional medicine through workshops, symposia, visit of experts, exchange of literature, etc.;
III. To pursue harmonized approach for the education, practice, research, documentation and regulation of traditional medicine and involvement of traditional medicine practitioners in health services;
IV. To explore the possibility of promoting mutual recognition of educational qualifications awarded by recognized Universities, pharmacopoeias, monographs and relevant databases of traditional medicine;

V. To encourage development of common reference documents of traditional medicine for South East Asian countries;
VI. To develop regional cooperation for training and capacity building of traditional medicine experts;
VII. To encourage sustainable development and resource augmentation of medicinal plants in the South East Asian regional countries;
VIII. To establish regional centers as required for capacity building and networking in the areas of traditional medicine and medicinal plants and
IX. To exchange views, experiences and experts for integration of traditional medicine into national health systems in accordance with national policies and regulations.

Source: http://www.pib.nic.in/newsite/erelease.aspx?relid=92213.

List 4. LIST OF INDIAN MEDICINAL PLANTS IN CITES

Appendix I:

1. *Saussurea costus*

Appendix II:

2. *Aquilaria malaccensis*
3. *Cycas bedomii*
4. *Dioscorea deltoidea*
5. *Rauvolfia serpentina*
6. *Cibotium barometz*
7. *Podophyllum hexandrum*
8. *Pterocarpus santalinus*
9. *Nardostachys grandiflora*
10. *Nepenthes khasiana*
11. *Picrorhiza kurrooa*
12. *Taxus wallichiana*

Source: http://envis.frlht.org/cites.php.

List 5. RED LISTED INDIAN MEDICINAL PLANTS

1. *Acorus calamus* L.
2. *Adenia hondala* (Gaertn.) de Willde
3. *Aegle marmelos* (L.) Correa
4. *Baliospermum montanum* (Willd.) Muell.-Arg.
5. *Canarium strictum* Roxb.
6. *Celastrus paniculatus* Willd.
7. *Coscinium fenestratum* (Gaertn.) Colebr.
8. *Embelia ribes* Burm. f.
9. *Gloriosa superba* L.
10. *Holostemma ada-kodien* Schult.
11. *Strobilanthes ciliatus* Nees

12. *Oroxylum indicum* (L.) Benth. ex Kurz
13. *Piper longum* L.
14. *Pseudarthria viscida* (L.) Wight & Arn.
15. *Pterocarpus santalinus* L.f.
16. *Rauvolfia serpentina* (L.) Benth. ex Kurz
17. *Salacia fruticosa* Heyne ex Lawson
18. *Saraca asoca* (Roxb.) de Wilde
19. *Santalum album* L.
20. *Symplocos cochinchinensis* (Lour.) Moore ssp. laurina (Retz.) Nooteb.
21. *Terminalia cuneata* Roth
22. *Tinospora cordifolia* (Willd.) Miers.

Source: www.smpbkerala.org/red-listed-medicinal-plants.html.

List 6. MEDICINAL PLANTS SPECIES IN HIGH TRADE SOURCED FROM TEMPERATE FORESTS

1. *Abies spectabilis*
2. *Aconitum ferox*
3. *Aconitum heterophyllum*
4. *Berberis aristata*
5. *Bergenia ciliata*
6. *Cedrus deodara*
7. *Cinnamomum tamala*
8. *Ephedra gerardiana*
9. *Juniperus communis*
10. *Jurinea macrocephala*
11. *Nardostachys grandiflora*
12. *Onosma hispidum*
13. *Parmelia perlata*
14. *Picrorhiza kurroa*
15. *Pistacia integerrima*
16. *Rheum australe*
17. *Rhododendron anthopogon*
18. *Swertia chirayita*
19. *Taxus wallichiana*
20. *Valeriana jatamansi*
21. *Viola pilosa*

Source: NMPB.

List 7. 70 MEDICINAL PLANTS IN HIGH TRADE SOURCED FROM TROPICAL FORESTS

1. *Acacia catechu*
2. *Acacia nilotica*
3. *Acacia sinuata*

4. *Aegle marmelos*
5. *Albizzia amara*
6. *Alstonia scholaris*
7. *Anogeissus latifolia*
8. *Asparagus racemosus*
9. *Baliospermum montanum*
10. *Bombax ceiba*
11. *Boswellia serrata*
12. *Buchnania lanzan*
13. *Butea monosperma*
14. *Careya arborea*
15. *Cassia fistula*
16. *Celastrus paniculatus*
17. *Chlorophytum tuberosum*
18. *Cinnamomum sulphuratum*
19. *Clerodendrum phlomides*
20. *Coscinium fenestratum*
21. *Cyclea peltata*
22. *Decalepis hamiltonii*
23. *Desmodium gangeticum*
24. *Embelia tsjerium-cottam*
25. *Emblica officinalis*
26. *Garcinia indica*
27. *Gardenia resinifera*
28. *Gmelina arborea*
29. *Gymnema sylvestre*
30. *Helicteres isora*
31. *Holarrhena pubescens*
32. *Holoptelea integrifolia*
33. *Holostemma ada-kodien*
34. *Ipomoea mauritiana*
35. *Ixora coccinea*
36. *Lannea coromandelica*
37. *Litsea glutinosa*
38. *Lobelia nicotianaefolia*
39. *Madhuca indica*
40. *Messua ferrea*
41. *Mimusops elengi*
42. *Morinda pubescens*
43. *Mucuna puriens*
44. *Nilgirianthus ciliatus*
45. *Operculina turpethum*
46. *Oroxylum indicum*
47. *Premna serratifolia*
48. *Pterocarpus marsupium*
49. *Pterocarpus santalinus*

50. *Rauvolfia serpentina*
51. *Rubia cordifolia*
52. *Santalum album*
53. *Sapindus mukorossi*
54. *Saraca asoca*
55. *Schrebera swietenioides*
56. *Semecarpus anacardium*
57. *Shorea robusta*
58. *Smilax glabra*
59. *Soymida febrifuga*
60. *Sterculia urens*
61. *Stereospermum chelonoides*
62. *Strychnos nux-vomica*
63. *Strychnos potatorum*
64. *Symplocos racemosus*
65. *Terminalia arjuna*
66. *Terminalia bellirica*
67. *Terminalia chebula*
68. *Vateria indica*
69. *Wrightia tinctoria*
70. *Ziziphus xylocarpus*

Source: NMPB.

List 8. 36 MEDICINAL PLANTS IN HIGH TRADE SOURCED FROM CULTIVATION

1. *Abelmoschus moschatus*
2. *Acorus calamus*
3. *Adhatoda zeylanica*
4. *Aloe barbedensis*
5. *Alpinia calcarata*
6. *Azadirachta indica*
7. *Caesalpinia sappan*
8. *Cassia angustifolia*
9. *Catharanthus roseus*
10. *Cichorium intybus*
11. *Croton tiglium*
12. *Curcuma angustifolia*
13. *Curcuma zerumbet*
14. *Ficus benghalensis*
15. *Ficus religiosa*
16. *Gloriosa superba*
17. *Indigofera tinctoria*
18. *Inula racemosa*
19. *Jatropha curcas*

20. *Kaempferia galanga*
21. *Lawsonia inermis*
22. *Lepidium sativum*
23. *Ocimum basilicum*
24. *Ocimum tenuiflorum*
25. *Piper longum*
26. *Plantago ovata*
27. *Plectranthus barbatus*
28. *Pongamia pinnata*
29. *Prunus armeniaca*
30. *Saussurea costus*
31. *Silybum marianum*
32. *Simmondsia chinensis*
33. *Trachyspermum ammi*
34. *Vitex negundo*
35. *Withania somnifera*
36. *Ziziphus jujuba*

Source: NMPB.

List 9. MEDICINAL PLANTS SPECIES IN HIGH TRADE SOURCED LARGELY THROUGH FORESTS

1. *Aquilaria agallocha*
2. *Commiphora wightii*
3. *Glycyrrhiza glabra*
4. *Piper chaba*
5. *Quercus infectoria*

Source: NMPB.

List 10. 46 MEDICINAL PLANTS SPECIES IN HIGH TRADE SOURCED MAINLY FROM WASTELANDS

1. *Abrus precatorius*
2. *Achyranthes aspera*
3. *Aerva lanata*
4. *Andrographis paniculata*
5. *Bacopa monnieri*
6. *Boerhavia diffusa*
7. *Cardiospermum halicacabum*
8. *Cassia absus*
9. *Cassia tora*
10. *Centella asiatica*
11. *Centratherum anthelminticum*
12. *Citrullus colocynthis*
13. *Convolvulus microphyllus*

14. *Curculigo orchioides*
15. *Cynodon dactylon*
16. *Cyperus esculentus*
17. *Cyperus rotundus*
18. *Datura metel*
19. *Eclipta prostrata*
20. *Fumaria indica*
21. *Hedyotis corymbosa*
22. *Hemidesmus indicus*
23. *Hygrophylla schulli*
24. *Ipomoea nil*
25. *Merremia tridentata*
26. *Ocimum americanum*
27. *Peganum harmala*
28. *Phyllanthus amarus*
29. *Pluchea lanceolata*
30. *Plambago zeylanica*
31. *Pseudarthia viscida*
32. *Psoralea corylifolia*
33. *Sida rhombifolia*
34. *Sisymbrium irio*
35. *Solanum anguivi*
36. *Solanum nigrum*
37. *Solanum virginianum*
38. *Sphaeranthus indicus*
39. *Tephrosia purpurea*
40. *Tinospora cordifolia*
41. *Tragia involucrata*
42. *Tribulus terrestris*
43. *Trichosanthes cucumerina*
44. *Vetiveria zizanioides*
45. *Withania coagulens*
46. *Woodfordia fruticosa*

List 11. ITEMS BANNED FOR EXPORTS FROM INDIA (APPENDIX 2-LIST OF FLORA IN APPENDIX I (PROHIBITED SPECIES) & APPENDIX (ENDANGERED SPECIES) II OF CITES

1. Baddomes cycad (*Cycas beddomei*)
2. Blue Vanda (*Vanda coerulea*)
3. *Saussurea costus*
4. Ladies slipper orcid (Paphiopedilium species)
5. Pitcher plant (*Nepenthes khasiana*)
6. Red Vanda (*Renanthera imschootiana*)
7. Sarpagandha (*Rauvolifia serpentina*)
8. Ceropegia species

9. Shindal Mankundi (*Frerea indica*)
10. *Podophyllum hexandurm* (*emodi*) (Indian Podophyllum)
11. Cyatheaceae species (Tree Ferns)
12. Cycadacea species (Cycads)
13. Dioscorea deltoidea (Elephant's foot)
14. Euphorbia species (Euphorbias)
15. Orchidaceae species (Orchids)
16. *Pterocarpus santalinus* (Redsanders)
17. Common Yew or Birmi Leaves (*Taxus wallichiana*)
18. Agarwood (Aquilaria malaccensis)
19. Aconitum species
20. Coptis teeta
21. Calumba wood (*Coscinium fenestrum*)
22. *Dactylorhiza hatagirea*
23. Kuru, Kutki (*Gentiana kurroo*)
24. Gnetum species
25. *Kampheria galanga*
26. *Nardostachys grandiflora*
27. *Panax pseudoginseng*
28. *Picrorhiza kurrooa*
29. Chirayata (*Swertia chirata*)

ANNEXURES

Annexure 1. REFERENCES ON CLINICAL TRIALS OF SINGLE HERBS USED IN AYURVEDA

Apamarga (Achyranthes aspera L.)

Shanker A, Parsai MR, Naqvi SMA, Jain JP. A clinical trial of *Apamarga* (*Achyranthes aspera*) in cases of *Shoth* (General Anasarca). *J Res Ayu Sidd* 1980; 1: 514–28.

Aswagandha (Withania somnifera Dunal)

Bikshapathi T, Kumari K. Clinical evaluation of *Ashvagandha* in the management of *Amavata. J Res Ayu Sidd* 1999; 19: 46–53.

Kuppurajan K, Rajagopalan SS, Sitaraman R, Rajagopalan V, Revathi R, Venkataraghavan S. Effect of *Aswagandha* (*Withania somnifera* Dunal) on the process of ageing in human volunteers. *J Res Ayu Sidd* 1980; 1: 247–58.

Aragvadha (Cassia fistula L.)

Das S, Sarkar PK, Sengupta A, Chattopadhyay A. A clinical study of *Aragvadha* (*Cassia fistula* L.) on *vicharchika* (eczema). *J Res Edu Indian Med* 2008; 14: 27–32.

Nair BKH, Nair CPR, Ramiah N, Kurup PB, Chandramouli K, Chandralekha B, Pillai KGB, Pai KN. *Cassia fistula* in pyoderma – a clinical trial. *J Res Indian Med* 1977; 12: 16–21.

Arjuna (Terminalia arjuna)

Jain V, Poonia A, Agarwal RP, Panwar RB, Kochar DK, Mishra SN. Effect of *Terminalia arjuna* in patients of angina pectoris (a clinical trial). *Indian Med Gaz* (New Series) 1992; 36: 56.

Tripathi VK, Singh B, Jha RN, Pandey VB, Udupa KN. Studies of *Arjuna* in Coronary Heart Disease. *J Res Ayur Sidd* 2000; 21: 37–41.

Chandra Sekhara Rao B, Singh RH, Tripathi K. Effect of *Terminalia arjuna* W. & A. on regression of LVH in hypertensives – a clinical study. *J Res Ayur Sidd* 2001; 22: 216–27.

Arka (Calotropis procera)

Jain PK, Verma R, Kumar N, Kumar A. Clinical trial of *Arka-mula-twak*-bark of *Calotropis procera* Ait. (R.Br.) on *atisar* and *pravahika* – a preliminary study. *J Res Ayur Sidd* 1985; 6: 88–91.

Bhallataka (Semecarpus anacardium L.)

Jha SD, Pandey VN. Clinical trial of *suddha bhallataka* on *gridhrasi. J Res Ayur Sidd* 1986; 7: 158–170.

Bharangi (Clerodendron serratum L.)

Shanker A, Parsal MR, Naqvi SMA, Jain JP. A clinical trial of Bharangi in cases of *Tamaka Swasa* (Bronchial Asthma). *J Res Ayu Sidd* 1980; 1: 470–8.

Gojivha (*Onosma bracteatum*)

Tripathi B, Mishra D, Dubey SD, Pandey LK, Tripathi KK. *Gojivha*: a clinical trial for its haemostatic action. *Light Ayur* 2009; 8: 1.

Guduchi (*Tinospora cordifolia* (Willd.) Hook.f. et Thoms.)

Gulati OD, Shah CP, Kanani RC, Pandya DC, Shah DS. Clinical trial of Tinospora in rheumatoid arthritis. *Rheumatism* 1980; 15: 143–8.

Japa (*Hibiscus rosa-sinensis* L.)

Tiwari PV. Preliminary clinical trial on flowers of *Hibiscus rosa-sinensis* as an oral contraceptive agent. *J Res Indian Med* 1974; 9: 96–8.

Jayapal (*Croton tiglium* L.)

Agarwal S, Deshpande PS, Shinde S, Gupta SS. Clinical trial of *Croton tiglium* and its alkaloid in alopecia areata. *Indian J Pharmacol* 1981; 13: 62.

Jyotishmati (*Celastrus paniculatus* Willd.)

Baranwal S, Gupta S, Singh RH. Controlled clinical trial of *Jyotishmati* (*Celastrus paniculatus* Willd.) in cases of depressive illness. *J Res Ayur Sidd* 2001; 22: 35–47.

Kalmegha (*Andrographis paniculata* Nees.)

Gupta SK, Rai NP. *Kalmegha* (*Andrographis paniculata*) and *kamala roga* (infective hepatitis) – a clinical trial. *Sach Ayur* 1999; 51: 544–6.

Kantakari (*Solanum xanthocarpum* L.)

Bector NP, Puri AS. *Solanum xanthocarpum* (*Kantakari*) in chronic bronchitis, bronchial asthma and non-specific unproductive cough (An experimental and clinical co-relation). *J Assoc Phys India* 1971; 19: 741–4.

Govindan S, Viswanathan S, Vijayasekaran V, Alagappan R. A pilot study on the clinical efficacy *Solanum xanthocarpum* and *Solanum trilobatum* in bronchial asthma. *J Ethnopharmacol* 1999; 66: 205–10.

Gupta PP, Dubey SD, Mishra JK, Ojha JK. A comparative study of *Brihati* and *Kantakari* in shvasa and kasa. *J Res Ayur Sidd* 1999; 20: 191–4.

Gupta SS, Verma SCL, Singh CM. Chemical and pharmacological studies on *Solanum xanthocarpum* (*kantakari*) in chronic bronchitis. *Indian J Med Res* 55(7): 723–732.

Jain JP. A clinical trial of *Kantakari* (*Solanum xanthocarpum*) in cases of *Kasa Roga*. *J Res Ayu Sidd* 1980; 1: 25–34.

Jain JP. A clinical trial of *Kantakari* (*Solanum xanthocarpum*) in cases of *Tamak Swasa* (Some respiratory diseases). *J Res Ayu Sidd* 1980; 1: 447–60.

Patil A, Arbar A, Deepti AK. Evaluations of bronchodilator effect of *Kantkari* extract nebulisation in wheezing children: A randomised clinical study. *Int J Res Ayur Pharm* 2014; 5: 295–8.

Shaw BP, Bera B. Treatment of tropical pulmonary eosinophilia with *Kantakari* (*Solanum xanthocarpum*) churna. *Deerghayu Int* 1985; 2: 3–8.

Karvir (*Nerium odorum* L.)

Singh P, Padhi MM, Tewari NS. A comparative clinical evaluation of *Karvir* in the management of *Pama* (Scabies). *J Res Ayu Sidd* 2002; 23: 51–5.

Lata Karanja (*Caeslpinia crista* L.)

Panda PK. Clinical study of *Caeslpinia crista* Linn. (*Lata Karanja*) in Malaria Patients. *J Res Ayur Sidd* 1998; 19: 122–7.

Mandukaparni (*Centella asiatica* L.)

Sharma AK, Sharma CM, Sharma UK. Clinical evaluation of Medhya Rasayana effect of *Mandukaparni* (*Centella asiatica*) – A scientific study. *J Res Ayu Sidd* 2005; 26: 32–44.

Mustaka (*Cyperus rotundus* L.)

Awasthi PK, Singhal KC, Singh N. *Cyperus rotundus* (Motha) – A clinical trial in some forms of arthritis. *Indian J Pharmacol* 1995; 27: 57 (abstract no. 64).

Saxena RC. *Cyperus rotundus* in conjunctivitis. *J Res Ayu Sidd* 1980; 1: 115–20.

Nirgundi (*Vitex negundo* L.)

Mishra SS, Dash NC, Das BK. A clinical study on *Gridhrasi* (Sciatica) and its management with *Nirgundi*. *J Res Ayu Sidd* 2003; 24: 42–50.

Palasha (*Butea superba*)

Cherdshewasart W, Nimsakul N. Clinical trial of *Butea superba*, an alternative herbal treatment for erectile dysfunction. *Asian J Androl* 2003; 5: 243–6.

Prajapati ML, Sharma AK, Sastry CHS. Clinical evaluation of *Palasha Pushpadi Churna* in the management of *Madhumeha Roga* (Diabetes Mellitus). *J Res Ayur Sidd* 2000; 21: 1–10.

Parijatha (*Nyctanthes arbor-tristis* L.)

Rao MK. A clinical trial of *parijatha* (*Nyctanthes arbor-tristis*) in *gridhrasi*. *Rheumatism* 1986; 21: 115–21.

Punarnava (*Boerhaavia diffusa* L.)

Singh RH, Udupa KN. Studies on the Indian indigenous drug *Punarnava* (*Boerhaavia diffusa* Linn.), part IV: preliminary controlled clinical trial in nephrotic syndrome. *J Res Indian Med* 1972; 7: 28–33.

Sarpagandha (*Rauwolfia serpentina*)

Vakil RJ. A clinical trial of *Rauwolfia serpentina* in essential hypertension. *British Heart J* 1949; 11: 350–55.

Shatavari (*Asparagus racemosus* L.)

Kishore P, Pandey PN, Pandey SN, Dash S. Treatment of duodenal ulcer with *Asparagus racemosus* Linn. (Preliminary Report). *J Res Ayu Sidd* 1980; 1: 409–16.

Shati (Hedychium spicatum Buch.-Ham.)

Sahu RB. Clinical trial of *Hedychium spicatum* in tropical pulmonary eosinophilia. *J Nepal Pharm Assoc* 1979; 7: 65–72.

Shigru (*Moringa pterygosperma*)

Hossain SMA, Hoque KMHS, Rashid A. Clinical trial: *Moringa pterygosperma* in the treatment of hypertension. *Chest Heart Bull* 1986; 10: 1–14.

Sirisa (*Albizzia lebbeck*)

Swam GK, Bhattathiri PPN, Rao PV, Acharya MV, Bikshapathi T. Clinical evaluation of *Sirisa Twak Kvatha* in the management of *Tamaka Shwasa* (Bronchial Asthama). *J Res Ayur Sidd* 1997; 18: 21–7.

Sunthi (*Zingiber officinale* Rosc)

Nanda GC, Tekari NS, Kishore P. Clinical evaluation of *Sunthi* (*Zingiber officinale*) in the treatment of *Grahni Roga*. *J Res Ayur Sidd* 1993; 14: 34–44.

Tamalapatra (*Cinnamonum tamala* L.)

Chandola HM, Tripathi SN, Udupa KN. Effect of *C. tamala* on plasma insulin vis-à-vis blood sugar in patients of diabetes mellitus. *J Res Ayu Sidd* 1980; 1: 345–57.

Tulsi (*Ocimum sanctum* L.)

Das SK, Chandra A, Agarwal SS, Singh N. *Ocimum sanctum* (tulsi) in the treatment of viral encephalitis (A preliminary clinical trial). *Antiseptic* 1983; 80: 323–7.

Vaca (*Acorus calamus* L.)

Mamgain P, Singh RH. Controlled clinical trial of the *lekhaniya* drug *Vaca* (*Acorus calamus*) in cases of ischaemic heart diseases. *J Res Ayur Sidd* 1994; 15: 35–51.
Rajasekharan S, Srivastava TN. Ethno-botanical study on *vacha* and a preliminary clinical trial on bronchial asthma. *J Res Indian Med* 1977; 12: 92–6.

Vibhituka (*Terminalia bellerica* Roxb.)

Trivedi VP, Nesamany S, Sharma VK. A clinical study of the anti-tussive and anti-asthmatic effects of *Vibhitakaphal churna* (*Terminalia bellerica* Roxb.) in the cases of *Kasa – swasa*. *J Res Ayur Sidd* 1982; 3: 1–8.

Yashtimadhu (*Glycyrrhiza glabra* L.)

Rai AN, Bhatia RPS, Chaudhary RC, Deshpande PJ. Effect of *Yashtimadhu* (*Glycyrrhiza glabra* Linn.) on conjunctivitis. *J Res Ayu Sidd* 1980; 1: 21–4.
Saxena RC, Gupta RN, Gupta GP, Bhargava KP. A clinical trial of *Glycyrrhiza glabra* in pemphigus. *J Indian Med Prof* 1965; 12: 5575–6.

Annexure 2: REFERENCES ON *TRIPHALA*

Pharmacognosy

Ashok Kumar D. Pharmacognostical investigations on *Triphala churnam*. *Anc Sci Life* 2007; 26: 40–4.
Bahulikar AS, Kashalkar RV, Pundlik MD. Visible Spectrophotometry in standardisation of herbal drugs – *Triphala Churna*. *Bull Medico-Ethno-Bot Res* 2002; 22: 118–27.
Desai AC, Savardekar MS. Specifications for Ayurvedic formulation *Triphala Churna*. 4th World Ayurveda Congress and Arogya Expo proceedings, pp. 264, 9–13 December 2010, Bengaluru, Karnataka, India.
Krishnamurthy KH. *Triphala* drugs of Ayurveda. *Indian J Pharm* 1968; 30: 297 (Abstract).
Krishnamurthy KH. Botanical aspects of the *Triphala* drugs of Ayurveda. *J Res Indian Med* 1970; 5: 95–105.
Lalla JK, Hamrapurkar PD, Mamania HM. Mineral content and microbial impurity of *Triphala churna* and its raw materials. *Indian J Trad Knowl* 2004; 3: 86–91.
Pawar V, Lahorkar P, Anantha Narayana DB. Development of a RP-HPLC method for analysis of *Triphala Churna* and its applicability to test variations in *Triphala Churna* preparations. *Indian J Pharm Sci* 2009; 71: 382–6.
Singh DP, Govindarajan R, Rawat AKS. High-performance liquid chromatography as a tool for the chemical standardisation of *Triphala* – an Ayurvedic formulation. *Phytochem Anal* 2008; 19: 164–8.
Thomas PJ, Satakopan S, Thakkar HJ, Mehta RC. Evaluation of *Triphala Churna*, part I: Microscopy, part II: Chemistry. *Indian J Pharm* 1974; 36: 172.

Pharmacology

Deep G, Dhiman M, Rao AR, Kale RK. Chemopreventive potential of *Triphala* (a composite Indian drug) on benzo (a) pyrene induced forestomach tumorigenesis in murine tumor model system. *J Exp Clin Cancer Res* 2005; 24: 555–63.
Dhanalakshmi S, Devi RS, Srikumar R, Manikandan S, Thangaraj R. Protective effect of *Triphala* on cold stress-induced behavioral and biochemical abnormalities in rats. *Yakugaku Zasshi* 2007; 127: 1863–7.
Gaind KN, Mital HC, Khanna SR. Anthelmintic activity of *Triphala*. *Indian J Pharm* 1964; 26: 106–7.

Ghosh D, Thejomoorthy P, Veluchamy G. Anti-inflammatory, anti-arthritic and analgesic activities of *Triphala*. *J Res Ayur Sidd* 1989; 10: 168–74.

Gupta SK, Kalaiselvan V, Srivastava S, Agrawal SS, Saxena R. Evaluation of anticataract potential of *Triphala* in selenite-induced cataract: *in vitro* and *in vivo* studies. *J Ayur Integr Med* 2010; 1: 280–6.

Inamdar MC, Rajarama Rao MR, Siddiqi HH. Purgative activity of *Triphala*. *Indian J Pharm* 1962; 24: 87–8.

Jagetia GC, Malagi KJ, Baliga MS, Venkatesh P, Veruva RR. *Triphala*, an Ayurvedic drug, protects mice against radiation-induced lethality by free-radical scavenging. *J Altern Complement Med* 2004; 10: 971–8.

Kalaiselvan S, Rasool MK. The anti-inflammatory effect of *Triphala* in arthritic-induced rats. *Pharm Biol* 2015; 53: 51–60.

Kaur S, Arora S, Kaur K, Kumar S. The *in vitro* antimutagenic activity of trip *Triphala* hala – an Indian herbal drug. *Food Chem Toxicol* 2002; 40: 527–34.

Kaur S, Michael H, Arora S, Härkönen PL, Kumar S. The *in vitro* cytotoxic and apoptotic activity of *Triphala* – an Indian herbal drug. *J Ethnopharmacol* 2005; 97: 15–20.

Mehta BK, Shitut S, Wankhade H. *In vitro* antimicrobial efficacy of *Triphala*. *Fitoterapia* 1993; 64: 371–2.

Naik GH, Priyadarsini KI, Mohan H. Free radical scavenging reactions and phytochemical analysis of *Triphala*, an ayurvedic formulation. *Curr Sci* 2006; 90: 1100–5.

Nariya MB, Shukla VJ, Ravishankar B, Jain SM. Comparison of gastroprotective effects of *Triphala* formulations on stress-induced ulcer in rats. *Indian J Pharm Sci* 2011; 73: 682–7.

Nariya MK, Shukla V, Jain S, Ravishankar B. Comparison of enteroprotective efficacy of tri *Triphala* formulations (Indian herbal drug) on methotrexate-induced small intestinal damage in rats. *Phytother Res* 2009; 23: 1092–8.

Ponnusankar S, Pandit S, Babu R, Bandyopadhyay A, Mukherjee PK. Cytochrome P450 inhibitory potential of *Triphala* – a Rasayana from Ayurveda. *J Ethnopharmacol* 2011; 133: 120–5.

Rajpal M, Sharma JN. Pharmacological study of alcoholic extract of *Triphala* and its anti-inflammatory activity in combination with petroleum ether extract of gum guggulu. *Indian J Pharmacol* 1984; 16: 99.

Ramasundaram SK, Narayanaperumal JP, Rathinasamy SD. Immunomodulatory activity of *Triphala* on neutrophil functions. *Biol Pharm Bull* 2005; 28: 1398–1403.

Rasool M, Sabina EP. Anti-inflammatory effect of the Indian Ayurvedic formulation *Triphala* on adjuvant-induced arthritis in mice. *Phytother Res* 2007; 21: 889–94.

Sabina EP, Rasool M. An *in vivo* and *in vitro* potential of Indian Ayurvedic herbal formulation *Triphala* on experimental gouty arthritis in mice. *Vasc Pharmacol* 2008; 48: 14–20.

Saravanan S, Srikumar R, Manikandan S, Jeya Parthasarathy N, Sheela Devi R. Hypolipidemic effect of *Triphala* in experimentally induced hypercholesteremic rats. *Yakugaku Zasshi. J Pharm Soc Japan* 2007; 127: 385–88.

Sharma JN, Rajpal MN, Rao TS, Gupta SK. Some pharmacological investigations on the alcoholic extract of *Triphala* alone and in combination with petroleum ether extract of oleo gum resin of *Commiphora mukul*. *Indian Drugs* 1988; 25: 220–3.

Shivaprasad HN, Kharya MD, Rana AC. Antioxidant and adaptogenic effect of an herbal preparation, *Triphala*. *J Nat Rem* 2008; 8: 1.

Sowmya SR, Antony S. Hypoglycemic effect of *Triphala* on selected non-insulin dependent diabetes mellitus subjects. *Ancient Sci Life* 2008; 17: 45–9.

Srikumar R, Parthasarathy NJ, Shankar EM, Manikandan S, Vijayakumar R, Thangaraj R, Vijayananth K, Sheeladevi R, Rao UA. Evaluation of the growth inhibitory activities of *Triphala* against common bacterial isolates from HIV infected patients. *Phytother Res* 2007; 21: 476–80.

Sujata N, Kumar S, Gupta GD, Rai NP. Hepatoprotective effect of *Triphala* in infective hepatitis (hepatitis B) and an experimental study. *AYU* 2008; 29: 176–80.

Vani T, Rajani M, Sarkar S, Shishoo CJ. Antioxidant properties of the ayurvedic formulation *Triphala* and its constituents. *Int J Pharmacog* 1997; 35: 313–7.

Clinical study

Maurya DK, Mittal N, Sharma KR, Nath G. Role of *Triphala* in the management of pyorrhoea. *Sach Ayur* 1995; 48: 390–1.

Maurya DK, Mittal N, Sharma KR, Nath G. Role of *Triphala* in the management of periodontal disease. *Phytother Res* 1997; 16: 91–3.

Mukherjee PK, RAI S, Bhattacharyya, Debnath PK, Biswas TK, Jana U, Pandit S, Saha BP, Paul PK. Clinical study of '*Triphala*' – a well known phytomedicine from India. *Iranian J Pharmacol Therap* 2006; 5: 51–4.

Srinagesh J, Krishnappa P, Somanna SN. Antibacterial efficacy of *Triphala* against oral streptococci: an *in vivo* study. *Indian J Dent Res* 23: 696.

Sujata N, Kumar S, Gupta GD, Rai NP. Hepato-protective effect of *Triphala* in infective hepatitis (hepatitis B): A clinical and an experimental study. *AYU* 2008; 29: 176–80.

Tandon S, Gupta K, Rao S, Malagi KJ. Effect of *Triphala* mouthwash on the caries status. *Int J Ayurveda Res* 2010; 1: 93–9.

Annexure 3: REFERENCES ON NIMBIDIN

Gurpreet K, Alam M, Athar M. Nimbidin suppresses functions of macrophages and neutrophils: relevance to its antiinflammatory mechanisms. *Phytother Res* 2004; 18: 419–24.

Pillai BKR, Amma KCB, Nair SS, Pillai NGK, Nair CPR. The effect of *nimbathiktha* (nimbidin) in *kitibha* (psoriasis) – a double blind clinical study. *J Res Ayur Sidd* 2002; 23: 42–50.

Pillai N, Suganthan RD, Seshadri C, Santhakumari G. Anti-gastric ulcer activity of nimbidin. *Indian J Med Res* 1978; 68: 169–75.

Pillai NR, Santhakumari G. Anti-arthritic and anti-inflammatory actions of nimbidin. *Planta Med* 1981; 43: 59–63.

Pillai NR, Santhakumari G. Effects of nimbidin on acute and chronic gastro-duodenal ulcer models in experimental animals. *Planta Med* 1984; 50: 143–6.

Pillai NR, Santhakumari G. Toxicity studies on nimbidin, a potential antiulcer drug. *Planta Med* 1984b; 50: 146–8.

Pillai NR, Santhakumari G. Effect of nimbidin on gastric acid secretion. *Anc Sci Life* 1985; 5: 91–7.

Pillai NR, Suganthan D, Santhakumari G. Analgesic and antipyretic actions of nimbidin. *Bull Medico-Ethno-Bot Res* 1980; 1: 393–400.

Pillai NR, Santahkumari G, Laping J. Some pharmacological actions of nimbidin. *Anc Sci Life* 1984; 4: 88–95.

Rajasekharan S, Pillai NGKP, Kurup PB, Pillai KGB, Nair CPR. Effect of nimbidin in psoriasis – A case report. *J Res Ayur Sidd* 1980; 1: 52–8.

INDEX

A

Abhayarishta 100
Abhraka Bhasma 54, 99, 100, 214, 235
adverse drug reactions 41, 43, 44, 59, 75
Agnimukha churna 100
Agnitundi Rasa 101, 140, 159
Ajmodadi churna 101
Amalaki Rasayana 102, 160
Amalkadi Ghrita 102
Amlapitta Mishran 103
Anu taila 103, 131, 179
Apamarga Kshara 103, 104, 149, 183
Apamarga Kshara Tailam 104
Apasmarari rasa 108
Aravindasava 106
Aristolactams 55
Aristolochia bracteata 56
Aristolochia fangchi 56
Aristolochia indica 56
Aristolochic acids 55–57
Arjunarishta 104
Arka lavana 104
Arkadi Kvatha Churna 105
Arogyvardhini vati 105
Ashthamulika Taila 108
Ashwagandha 51
Ashwagandhadi lehyam 107
Ashwagandharishta 107
Avartita Panchtikta Ghritha 151
Avipattikar churna 108, 109
Ayurvedic Pharmacoepidemiology 75
Ayurvedic pharmacy 11, 67–69, 241, 242
Ayush Premium Mark 13, 77, 81–83, 86
AYUSH Research Portal 88
Ayush Standard Mark 13, 77, 81, 86

B

Balacaturbhadrika churna 109
Balachaturbhadra Avaleha 109
Balarishta 109
bhasma 30, 42, 46, 48, 49, 54, 99, 100, 122, 123, 126–128, 132, 136–138, 141–146, 153, 154, 157, 163, 173–175, 177, 178, 184, 185, 187, 189, 190, 214, 227, 228, 230–232, 234–237

Bilvadileha 110
Brahma Rasayana 111, 112
Brahmi Ghrita 112, 113
Brahmi vati 113

C

Candanasava 113, 114
Carbamazepine 180
Centre for Safety & Rational Use of Indian Systems of Medicine 45
Chandrakala rasa 114
Chandraprabha vati 53, 114, 115
Chitrakadi vati 117

D

Dasamula Taila 119
Dasamularishta 118
Dashanga Kwatha Ghana 117
Dashanga Yoga 117
Declaration of Alma-Ata 5
Dhanvantara Gutika 119
Dhatri Lauha 119, 120, 130
diclofenac sodium 118, 180
Draksarista 121
Drakshadi vati 120
Drakshasava 121
Drakshavaleha 120, 121
Drug Safety 41

E

Eladi Gutika 121, 122

G

Gandhaka Rasayana 106, 122
Godanti bhasma 122, 123, 127, 227, 228
Gokshuradi churna 124
Gokshuradi Guggulu 123
Gokshuradi Vati 123
Goksuradi Guggulu 123, 129
Good Agricultural Practices (GAP) 6
Good Clinical Practice 58, 59
Good Manufacturing Practices 6, 22, 26, 39, 48, 226

Guduchi Ghana 124
Guduchi ghrita 124

H

Hajrul Yahud Bhasma 126
Hara Lauha 163
Haridra Khanda 126, 127
Haridradi Ghrita 126
Haritaki Churna 189
Hartala bhasma 127, 228
Hartala godanti bhasma 127, 228
Hartaladi 141
herbal medicines 3–5, 37, 45, 46, 48, 55
Hinguadi taila 127
Hinguleshwar Rasa 127, 179
Hingusauvarchaladi ghrita 128

I

IMCC Act 1970 26, 65, 66
Indomethacin 163, 168, 180
Indrayanadi Yog 128
International Conference on Drug Regulatory Authorities 5
Isoniazid 180

J

Jatyadi Ghrita 128, 159
Jatyadi Taila 128, 129
Jeevantyadi Ghrita 129
Jwarhar mahakashay 129

K

Kajjali 54, 130, 137, 159
Kajjali Bhasma 54
Kalpita Yoga 130
Kamadudha Rasa 130, 133
Kanakasava 131
Kanchnar Guggulu 123, 129, 130
Kankayana vati 131, 132
Kantalauha Bhasma 137
Kapardika bhasma 187
Kaphaketu Rasa 103, 131
Karpura shilajit bhasma 132, 232
Kasishadi taila 131, 132
Keharuba Pisti 130, 132, 133
Khadirarista 133
Kshatantak Malam 132
Kshiramandura 133
Kushmandadi Ghrita 134
Kushta Tila Kalan 174
Kutajarishta 134, 135

L

Laghugangadhar Churna 135
Lakshadi Guggulu 136

Lasunadi vati 136
Lauha bhasma 54, 136, 137, 142
Lavangadi Vati 138
Lodhrasava 138, 155
Loknath rasa 138

M

Madhuhari churna 139
Maducasava 121
Mahalaxmivilas Ras 47, 54, 139
Mahalaxmivilas Rasa 139
Mahamrutyunjaya rasa 140
Maharasnadi Quatha 139, 187
Mahayograj Guggul 47, 54
Mahayograj guggulu 52, 101, 140, 159
Malla-Sindur 141
Manasamitra vatakam 141
Mandoor parpati 142
Mandura bhasma 54, 137, 141, 142
Mandura Vataka 142, 143
Manikya Bhasma 141, 236
Manjishtadi kashayam 143
Manjishthadi Ghrita 143
Manjisthadi churna 143
Mehamudgara vati 143, 144
Mukta bhasma 144
Mukta Shukti Bhasma 144, 236

N

Naga bhasma 49, 145, 146, 214
Narcha churna 146
Nardeeya Laxmi Vilas Rasa 122
Nardiyalaxmi Vilas Ras Mishran 147
Narsingh churna 100
National Accreditation Board for Certification Bodies 78, 80
National AYUSH Mission 64–66
National Center for Complementary and Alternative Medicine 41
Navajeevan 130, 133
Navayasa Lauha 120
Navbal Rasayan 147
Navratna rasa 52, 53, 147
Nidigdhikadi kvatha 148
Nirgundi Ghana Vati 148
Nisamalaki Churna 148
nishadi tail purana 149
Nityananda rasa 149
Nyagrodhadi churna 149

P

Padmapatradi yoga 149
Palasha Kshara Sutra 149
Palasha Pushpadi Churna 150
Pancaguna taila 150

Index

Pancha Nimba Churna Vati 150
Panchakola Avaleha 150
Panchamrita parpati 150
Panchatikta Ghrita Guggulu Vati 151
Panchavaktra rasa 151
Panchavalkal kwath 191
Panchvalkala 151
Pashanabhedadi Ghrita 152
pefloxacin 180
Phalagritam 152
Pharmacoepidemiology 75
pharmacopoeia monograph 4, 243
Pharmacovigilance 41–45
Pharmacovigilance of Ayurvedic Medicine 43, 44
Pharmacy Council of India 68, 69
phytomedicines 4, 56
Pinda Taila 152
Pippali churna 160
Pippaliyadi yoga 153
Pravala Bhasma 153
Puga Khanda 154
Punarnavasava 154
Punarnavashtak kwath 155
Pushpadhanwa Rasa 156
Pushyanuga churna 138, 155
Puskara Mooladi churna 155

R

Rajata Sindura 156
Raktapittantaka Louha 156
Ras Manikya Ras 54, 159
Rasa Aushadhies 48
Rasa Bhasma 157, 214
Rasa karpura 157, 229
Rasa Parpati 159, 233
Rasa pushpa 159, 230
Rasaka bhasma 157
Rasamanikya 158, 159
Rasnadi Guggulu 160
Rasnasaptaka Quatha 160
Rasona Pinda 161
Rauvolfia serpentina 57, 238, 244, 245, 247
Regulatory affairs 1, 3
Rifampicin 180
Rohitakarista 161

S

Safety Profile of Ayurvedic Dosage Forms 44
Sameera Pannaga Rasa 161
Samsharkara churna 162
Sanjivani vati 162
Saptamrita Lauha 162, 183
Saptavartita Hingusauvarchaladi ghrita 128
Saraswataristam 162, 163
Sarva-Juara 163
sastiputa naga bhasma 145, 146

Satavari Mandur 164
Schedule 'T' of the Drugs and Cosmetics Act, 1940 6
Shankhapani Rasa 163
Sheetamshu rasa 164, 165
Shila Sindoor 165
Shilajit 132, 165–170, 232
Shirishavaleha 170
Shivagutika 164
Shivaksharpachan Churna 171
Shringa Bhasam 165
Simhanada Guggulu 171
Sitopaladi Churna 172
Smriti Sagara rasa 54
Srngyadi Churna 172
Sudarshan Churna 173
Sunder Vati 176
Surya Shekhara Ras 176
Sutshekhar Rasa 173
Sveta parpati 172, 214
Swarna bhasma 173, 174, 230
Swarna makshika bhasma 174, 230, 235
Swarna Parpati 175, 233
Swarna vanga 175
Swasa Kuthara rasa 54

T

Talisadya Churna 176, 177
Tamra Bhasma 54, 177, 178, 214, 236
Tankana Bhasma 178
tarakeswara rasa 178
The Drugs and Magic Remedies 20
The Indian Council of Medical Research 90
The Pharmacovigilance Program of India 43
The United Nations Convention on Biological Diversity 4
The World Health Assembly 5, 243
the World Health Organization 5, 10, 114, 243
traditional medicine 4, 5, 9, 19, 21, 43, 45, 49, 56, 67, 70, 242–244
Tribhuvan kirti rasa 127, 179
Triguna Rasa sindhoora 179
Trikatrayadi Lauha 179
Trikatu 151, 179–182
Triphala churna 131, 132
Triphala eye drops 183
Triphala ghrita 182
Triphala guggul 183
Triphala Mashi 183, 184
Trivanga bhasma 184, 231

V

Vacha Dhatryadi Avaleha 184
Vaishvanar Churna 185
Vanga Bhasma 185, 214
Varatika bhasma 187, 228, 237

Vardhamana Pippali Rasayana 185
Varunadi loha 186
Varunshigru Ghan Vati 186
Vasant kusumakara 187
Vasavaleha 186
Vatagajankusa rasa 187
Vatari Guggulu 139, 187, 188
Vayasthapana Rasayana 188
Veerataru Kwatha 188
Vibhitakyadi churna 188
Vidangadi churna 188, 189

Vidarikandadi Yog 189
Virya Sthambhak Vati 189

Y

Yashada bhasma 189, 190
Yavaksharadi vati 191

Z

Zahr Mohra 191